A MANUAL OF AQUATIC FUNGI

A Manual of Aquatic Fungi

Chytridiomycetes and Oomycetes

R D Khulbe

2021

Daya Publishing House®

A Division of

Astral International Pvt. Ltd.

New Delhi – 110 002

First Published, 2001

Reprinted, 2021

ISBN 9788170352228 (HB)

Published by : **Daya Publishing House®**
A Division of
Astral International Pvt. Ltd.
– ISO 9001:2015 Certified Company –
4736/23, Ansari Road, Darya Ganj
New Delhi-110 002
Ph. 011-43549197, 23278134
E-mail: info@astralint.com
Website: www.astralint.com

Printed at : **Replika Press Pvt. Ltd.**

DEDICATED
TO

Professor K.S. Bhargava F.N.A.
(b. 19th october, 1918)

Dr. R. Dayal
F.P.S.I.; F.N.A.Sc.
Professor Ex.

Deptt. of Mycology and Plant Pathology
Institute of Agricultural Sciences
Banaras Hindu University
Varanasi -221005

Foreword

It gives me immense pleasure to write a foreword for this book, 'A Manual of Watermolds' by Prof. R.D. Khulbe. There is no dearth of books in Botany in the market, but monographs of this type are rare. Students in this group in India must be finding difficulty in acquiring update information on morphology and taxonomy of watermolds, as monographic studies comparable to Coker's (1923) Saprolegniaceae; Sparrow's (1960) Aquatic Phycomycetes; Scott's genus *Aphanomyces*; Johnson's genus *Achlya*; Semour' s *Saprolegnia*; and Karling's Iconographia Chytridiomycetarum (1977) considered milestones in Mycology are lacking.

This book is an attempt to fill the void and such a monograph on Watermolds (filamentous Chytridiomycetes and Oomycetes) is first of its type from India. Probably this may be the first update compilation of work carried out by Indian authors in this group of organisms by Dr. Khulbe in this new millenium.

I congratulate Prof. R.D. Khulbe, Department of Botany, Kumaun University, Nainital for this valuable contribution, which will no doubt, be of considerable benefit to the students, teachers, research scholars of advanced mycology, as well as those engaged in research in this area in India and abroad.

R. Dayal
Retd. Emeritus Professor

Preface

Members of Chytridiomycetes and Oomycetes are mostly aquatic and are popularly known as 'watermolds'. However, a considerable number of these species also occur in moist soils. These fungi possess ability of colonizing a variety of substrates, while a few of them invade living hosts, i.e., plants and animals. They vary in their occurrence, distribution, asexual and sexual reproductive structures in relation to ecological factors. All these characters are closely linked with their physiological adaptability and specialization.

These fungi contribute to the energy flow and productivity of an aquatic and semi-aquatic ecosystems by their active role in the utilization and biodeterioration of a variety of organic and inorganic materials. Physico-chemical factors of the ecosystems play important roles in the growth, multiplication, distribution and seasonal periodicity of the watermolds (Coker, 1923; Forbes, 1935; Milovtsova, 1942; Naumov, 1954; Perrott, 1960; Dick and Newby, 1961; Hughes, 1962; Dayal and Tandon, 1962; Roberts, 1963; Srivastava, 1967; Alabi, 1971; Khulbe and Bhargava, 1977; Manoharachary, 1981; Khulbe *et al.*, 1993.)

The taxonomy of the watermolds can not remain static, when rest is changing. The morphological variability of any species must ultimately be linked to the genetic control of these characters. A considerble amount of taxonomic studies have been carried out both in America and Europe by a number of investigators including Coker(1923), Hohnk (1932), Coker and Mathews (1937), Middleton (1943), Sparrow (1943 and 1960), Johnson (1956 and 1971), Scott (1961), and Seymour (1970).

In India, Butler (1907) was the first to report occurrence of *Allomyces* and *Pythium* species. Later, a large number of investigators have reported different species of Chytridiomycetes and Oomycetes from tropical, subtropical and temperate habitats. Their reports are widely scattered in different journals and theses, which are not easily available to the students and researchers of aquatic mycology. Dayal and Kiran (1988), however, have published a comprehensive book on zoosporic fungi of India, but they did not mention taxonomy of these fungi. In the present monograph, the author has tried to compile taxonomic accounts of all the filamentous forms of Chytridiomycetes (Blastocladiales and Monoblepharidales) and Oomycetes (Saprolegniales, Leptomitales and Lagenidiales) reported from different parts of India up to 1998. A general account of the distribution of the species found under different ecological conditions of India is also given. Keys to the different species of each genus are also described.

In this monograph the taxonomic variation of each species found under Indian conditions are given. The species in each genus are described in alphabetical order just after key, and the page number of each species is also given in the key.

The Chytridiomycetes and Oomycetes are separated on the basis of flagella, i.e., uniflagellate and biflagellate zoospores, respectively. These fungi are present in waters and moist soils, and can be isolated from different animals and plants using them as baits.

These species are mostly saprophytes, however, some are parasites as well. The vegetative bodies of Blastocladiales are eucarpic, stout, cylindrical, arboreal, sympodially or dichotomously branched, with or without rhizoidal systems and setae. The members of Monoblepharidales are characterized by filamentous, branched mycelium having constrictions at intervals with pseudosepta. The protoplasm is foamy and reticulate in nature. In both Blastocladiales and Monoblepharidales, asexual reproduction takes place by means of uniflagellate zoospores. The zoospores are pyriform or spherical with posterior whiplash flagellum, monomorphic and monoplanetic. Zoosporangia are pyriform, cylindrical or filamentous and papillate.

In Blastocladiales, thick walled resting sporangia or spores are also developed, while in Monoblepharidales sporangia are tapering and internally proliferated in most of the species.

It is again interesting to note that in Blastocladiales two types of zoospores are produced, i.e., one diploid, developed within thin walled zoosporangia, giving rise to sporothalli; while those produced by the thick walled sporangia are haploid and produce gametothalli. In sexual reproduction, male and female gametangia are borne on gametothalli. They seldom occur in nature. Female gametangia are large, colourless, terminal or lateral, and male gametangia are coloured. Gametes are motile, anisogamous. After fusion (sexual reproduction) planozygote is formed, which germinates to form the sporophytic plant. In this state the fungus occurs in nature.

In Oomycetes, the fungi occur in water and soil (cultivated and non-cultivated), on various plant and animal baits. Most of the species are saprophytic, and a few of them show parasitic association with aquatic animals, mainly fishes, arthropods, plant seedlings and roots.

The vegetative bodies are filamentous, eucarpic extensively branched and coenocytic. All these fungi produce biflagellate zoospores. Mode of liberation and behaviour of zoospores is different in different genera and sometimes in different species of the same genus. Gemmae—the thick walled perennating asexual bodies are also developed, and behave either as zoosporangia or as single spore. The zoospores are dimorphic, primary ones are pyriform with two anteriorly placed flagella, i.e., one tinsel and the other whiplash type. Secondary zoospores are formed by encysted primary zoospores, which are reniform and with two lateral, flagella, whiplash flagellum is posteriorly directed.

The swarming behaviour of the zoospores is very interesting and taxonomically an important character. In some genera (*Phythiopsis, Saprolegnia, Isoachlya and Leptolegnia*), they swim actively; while in the other, suppression of motility starts at the tip of the zoosporangium (*Achlya* and *Aphanomyces*) or within the zoosporangium (*Aplanes, Thraustotheca, Dictyuchus* and *Brevilegnia*). In *Geolegnia,* swarming stage is completely suppressed. Liberation of primary encysted zoospores from sporangia is also important, which come out through separate pores (*Dictyuchus*) or released in groups by disintegration or cracking (*Thraustotheca*) or dissolution (*Brevilegnia and Geolegnia*) of the walls.

Sexual reproduction is also interesting in these fungi, which is performed by gametangial copulation (oogamy) due to non-motile gametes. Oogonial and antheridial branches, their origin, shape, size, thickness, number etc., are of taxonomic values. Antheridia are of hypogynous, androgynous, monoclinous and diclinous type depending upon their origin from oogonial cell, oogonial stalk, same filament or other filament, respectively. After fertilization, non-motile oospores are formed in oogonia. Their number per oogonium, size, thickness, maturity period and arrangement of inner oil and protoplasmic granules are different in different genera, and species. In centric oospores, protoplasmic oil granules are arranged in one or two peripheral rows; in eccentric oospore, one large oil globule is located on one side; in subcentric oospore, one layer of small oil droplets on one side and two or more on the opposite side of the oospore; while in subeccentric oospore, one or more layers of large oil droplets on one side of the oospore (Seymour, 1970; Dayal and Kiran, 1988 and Khulbe, 1993).

In Leptomitales, mycelium is delicate, flocculent and branched sparingly. Hyphae are constricted into segments. Zoosporangia produce in apical and basipetal succession, escape as in *Saprolegnia.* Zoospores are monoplanetic and eccentric. Sexual reproduction is oogamous, single or many oospores are developed within oogonium.

The members of Lagenidiales (*Olpidiopsis*) are unicellular, holocarpic, endobiotic, obligate, parasitic in algae and fungi. The parasite occupies apical or intercalary position in swelling of the hosts. Entire thallus is converted into zoosporangium. Zoospores are minute, many, kidney shaped, laterally biflagellate as secondary zoospores of Saprolegniales, swim violently within zoosporangium, and liberate out from zoosporangium and host walls through long papillae. They then infect suitable hosts and make new infections. Here, sexual reproduction is oogamous, both oogonia and antheridia (companion cells) are present within the swellings of the hosts. Oogonia are smooth, tuberculate or spiny, spherical, elliptical; antheridial cell is small and tuberous. Single or two oospores are developed within oogonium, which are smooth, tuberculate or spiny, perennating or germinate on dissolution of the host and oogonial walls to give rise many zoospores.

For the isolation of these fungi, various plant and animal baits are used. Sterilized endocarps of hemp and *Brassica* seeds are most suitable baits. For the culture of these fungi various media (P.D.A., C.M.A., S.P.S., S.A., YyPs, YpG, Raper's medium, Dick's Agar Medium, *Brassica* Agar Medium etc.) are suitable. For the detailed methods of their isolation, purification, culture, identification etc., students are advised to consult Johnson (1956), Sparrow (1960), Seymour (1970), Dayal and Kiran (1988) and Khulbe (1993).

For the present endeavour, I am very much grateful to my teacher Dr. K. S. Bhargava, F.N.A. formerly Professor and Head, Department of Botany, Kumaun University, Nainital and later, Department of Botany, University of Gorakhpur, for his kind guidance, inspiration and blessings.

I am also grateful to Dr. R. Dayal, Retired Professor, B. H. U., Varanasi for critical examination of the manuscript and many suggestions.

Thanks are also due to Prof. R.S. Mehrotra, Kurukshetra University; Prof. G.P. Agarwal and Prof. S.K. Hasija, R.D. University, Jabalpur; Prof, R.S. Dwivedi, B.H.U., Varanasi; Prof., Rama Rao and Prof. Manoharachary, Osmania University, Hyderabad; Prof. B.P. Kapadnis, Poona University, Pune; Dr. P. N. Chowdhry, I.A.R.I., New Delhi; Dr. G.C. Srivastava, St. Andrews College, Gorakhpur; Dr. P.B. Awasthi, Bareilly College, Rohilkhand University, Bareilly; Dr. R. P. Mishra, R. D. University, Jabalpur; Dr.J. K. Misra, Lucknow University and Dr. S. K. Prabhuji, Padrauna, U. P. for sending their published research papers, photographs and suggestions from time to time to bring out this volume.

My thanks are also due to all the Heads, colleagues and research scholars of Botany Department, D. S. B. Campus, Kumaun University, Nainital who have made a strong platform for the work on aquatic mycopathology in Kumaun University. I cannot forget the help rendered by my family members, other research scholars, typist, cartographer and friends to make this endeavour successful.

I hope this monograph will be useful to scientists, teachers and research scholars. Suggestions and criticism are heartily invited, so that they may be included in the subsequent edition.

July, 2000
Nainital

R. D. Khulbe

Contents

I. CHYTRIDIOMYCETES

BLASTOCLADIALES

1. **Blastocladiaceae :**
   Allomyces
   Blastocladia
   Blastocladiella
2. **Catenariaceae :**
   Catenaria
   Catenomyces

Allomyces Butler

Ann. Bot. London 25 : 1023 – 1035, 1911
Aquatic Phycomycetes, p. 669, 1960

Key to the Species of Allomyces

Sex organs absent .... A
Sex organs present................................ B

A_1 Branching dichotomous, mature plant body pinkish.......................... *A. moniliformis,* p. 7
A_2 Branching sympodial, mature plant body colourless............... *A. anomalus,* p. 4
B_1 Secondary zoosporangia in long chain............................... *A. neo-moniliformis,* p. 7
B_2 Secondary zoosporangia not in chain...C
C_1 Male gametangia not coloured............ *A. javanicus,* p. 6
C_2 Male gametangia coloured.................. *D.*
D_1 Male gametangia hypogynous............ *A. arbuscula,* p. 5
D_2 Male gametangia lateral..................... *A. recurvus,* p. 8

Blastocladiaceae

Allomyces anomalus Emerson

Lloydia 4 : 133, 1941

(Plate 1, Figs. 1–8)

Principal hyphae stout, slender, crooked, sympodially branched, pseudo septate; 12–90 μm in diameter at base, mostly 15–20 μm, basal joints 50–500 μm long, central joints up to 750 μm long; tips blunt, hyaline; zoosporangia oval, 20–42 × 45–75 μm; zoospores escaping through an apical pore, uniflagellate, oval, monoplanetic; encysted zoospores 6–12.5 μm in diameter, mostly 10-5 μm; resting sporangia abundant, 30–62 μm × 18-42 μm, mostly 40-45 × 30-38 μm, produce zoospores after rest; sex organs absent (Khulbe, 1985).

The species was reported by Emerson (1941) from Pusa (Bihar), and Saksena and Rajgopalan (1958) from Sagar. Later, it was isolated from pond water of Vikarabad, Hyderabad (Manoharachary, 1977); temple pond water and mud at 6.2-7.8 pH, 3-12 ppm organic matter (Manoharachary, 1979b); temple pond of Antagiri Hills, Vikarabad and Typha pond, Osmania University Campus, Hyderabad during monsoon and winter seasons at 20.6 - 26.10°C and 5.9-7.9 pH (Manoharachary, 1979 a); soil and water samples around I.A.R.I., New Delhi during summer (March-June) at 30-45°C temperature (Chowdhry and Agarwal, 1980a), some soil and water samples of Rampur, Lucknow, Tejpur, Manglore, Haldwani and Kasargold (Chowdhry and Agarwal, 1980b); pond water, Hyderabad during monsoon from June to September (Manoharachary and Rao, 1981); temple pond of Antagiri Hill, Vikarabad and Alisagar lake, Nizamabad during monsoon (frequently), winters and summer (less frequently) at 6.2-7.6 pH (Manoharachary *et al.*, 1983); a forest soils of Haldwani, Ganai, Nainital, Pithoragarh and Ranikhet (Khulbe, 1985); sal forest soils of Haldwani during summer and rainy seasons and oak forest soils of Nainital in summer at 14-30°C temperature, 2-50 per cent moisture, 0.8-4.9 per cent organic carbon and 5.7-6.9 pH (Khulbe, 1991).

Gupta and Mehrotra (1989a, b & 1992) isolated it from Kurukshetra on mustard and *Sesamum* seed baits from Brahma Sarovar tank under moderate to high temperature conditions (21.4.-34.8°C).

Allomyces arbuscula Butler

Ann. Bot. London, *28* : 1027, 1911.

(Plate 1 ; Figs. 9-14)

Principal hyphae stout, dichotomously branched, pseudoseptate, 10-60 µm in diameter at base, mostly 14-18 µm, basal joints 80–600 µm long, central joints up to 700 µm long; tips blunt, hyaline; zoosporangia oval, hyaline, 23-50 × 50-80 µm; zoospores escaping singly through an apical pore, or through distinct discharge papillae, uniflagellate, oval, monoplanetic, encysted zoospores 6.5-13 µm in diameter, mostly 10.5 µm; resting sporangia abundant, 60-130 µm long by 50-100 µm in diameter, mostly 40-45 × 33-35 µm, produce into zoospores after a period of rest; sex organs abundantly in pairs; female gametangia terminal, spherical, hyaline, 40-100 µm, male gametangia hypogynous, orange coloured, barrel shaped, 30-70 µm in diameter (Khulbe, 1985).

This species was isolated from Poona (Butler, 1911), Bombay and Calcutta (Emerson, 1941); water and soils of Ramgarh Lake, University garden pond from Gorakhpur (Bhargava and Singh, 1965); Varanasi (Thakurji, 1967); scrub jungle and dry waste land soils of Hyderabad (Manoharachary, 1977), mud soils of Hyderabad in all the seasons at 30-57 per cent moisture, 7-8.2 pH, 12.71-22 per cent organic matter, 50-171 mg soil salts per 100g mud and 6.8-12.3 mg chlorides per 100g mud (Manoharachary, 1979a); mud soil, air dried soils from road side ditches, clay loam forest soils, sandy loam dry soil at 6.2-8.2 pH, 3-12 ppm to 14-24 per cent organic matter during summer, monsoon and winter with 3.7-33.3 per cent frequency (Manoharachary, 1979b); temple pond of Antagiri Hill, Vikarabad, Typha pond, Osmania University Campus, 6-7.9 pH (Manoharachary, 1979 a and b,); soil samples of Ayarpata and Tiffin Top, Naini Tal (Sati and Khulbe, 1980); soil of Delhi (Chowdhry and Agarwal, 1980 a and b); soil and water samples of Delhi, Rampur, Haldwani, Lucknow, Tezpur, Mangalore, Coonor, Karagod (Chowdhry and Agarwal, 1980b); soil and water of Pune (Kapadnis, 1980); pond A and B, Hyderabad during all the seasons (Manoharachary and Rao, 1981); muddy soil of Mannanur forest, Andhra Pradesh (Rao and

Manoharachary, 1981), two tanks of Nizamabad, A.P. (Bhairabnath and Manoharachary, 1985), humid soils and many water bodies (at pH 6.2-8.2), viz., temple pond of Antagiri Hill, Vikarabad, Alisagar lake, Nizamabad district (Manoharachary *et al.*, 1983), soils of Gorakhpur (Prabhuji, 1984); forest soils of Haldwani, Ganai, Naini Tal, Pithoragarh and Ranikhet (Khulbe, 1985); oak forest soils of Naina Peak, Naini Tal (Summer & rainy) and sal forest soils, Haldwani (all seasons) at 14-30°C temperature, 2-90 per cent moisture, 0.4-4.2 per cent organic carbon, 5.3-6.9 pH (Khulbe, 1985 & 1991). Manoharachary (1979a) considered this species primarily as an aquatic and secondarily terrestrial. Gupta and Mehrotra (1989 a, b and 1992) isolated it on mustard and *Sesamum* seed baits from Brahma Sarovar and Sanhit Sarovar, Kurukshetra under moderate to high temperature conditions (21.4-34.8°C).

Khare (1992) found it as most common species in Nakatia river, Jalalpur lake and Bareilly club pond of Bareilly.

Allomyces javanicus Kniep

Berichte Deutsch. Bot., Ges., **47** : 211, 1929.

(Plate 2; Fig. 1)

Thallus sympodially or dichotomously branched; basal cell variable in size; zoosporangia ovoid, apical or catenulate, 40-90 × 22-60 μm; with one to many papillae; zoospores ovoid, 10-12 × 8-10 μm; resting spores spherical, ovoid, 30-65 × 20-40 μm; exospores thick, brown, reddish, gametothalli similar to sporothalli; gametangia terminal in pairs, simple to catenulate, male gametangia terminal, 32 × 25 μm, female gametangia ovoid 50-55 × 30-35 μm.

This species was isolated from soils of Pusa and Darbhanga, Bihar (Emerson, 1941), Madras (Subramanium, 1952), Gorakhpur (Prabhuji and Srivastava, 1977a); air dried soils from road side ditches of Hyderabad and Vikarabad (pH 7.2-8.2, organic matter 3.1-6 per cent) during winter, summer and monsoon seasons (Manoharachary, 1979b); soil and water samples around IARI, New Delhi during summer and rainy seasons at 20-45°C (Chowdhry and Agarwal, 1980a).

Prabhuji and Srivastava (1977a) found coiling nature of the

vegetative hyphae, which were more pronounced in the renewed hyphal branches just below the resting sporangia. In gametothalli sometimes male gametangia were found shifted from their terminal portion to the lateral position in relation to female gametangia.

Gupta and Mehrotra (1989 a, b and 1992) isolated it on mustard and *Sesamum* seed baits from Brahma Sarovar, Kurukshetra at 14.2-34.8°C temperature.

Allomyces moniliformis Coker and Braxton

J. Elisha Mitchell Sci. Soc. **42**: 139, 1926

Mycologia **30** : 127, 1938

(Plate 2; Fig. 2)

Thallus dichotomously branched, basal cell 100-150 × 15-50 µm; old culture slightly pink; primary sporangia clavate, cylindrical, 50-140 × 15-35 µm with an apical papilla; secondary sporangia basipetal, spherical, elliptical, with one to four lateral papillae, pink, mature brown; zoospores ovoid, posteriorly uniflagellate, resting spores ovoid, 40-70 × 20-45 µm, exospores thick walled, orange brown and pitted.

This species was recorded from soils of Varanasi (Thakurji, 1967 and 1970); silty soils (pH 6.5, moisture 25 per cent and organic contents 4.2 per cent) around Ramgarh lake, Gorakhpur (Prabhuji and Srivastava, 1977a); as a saprophyte on insect exuvae, Pune, (Kapadnis, 1980); from soil and water samples of Delhi, Rampur, Haldwani and Kasargod (Chowdhry and Agarwal, 1980b). Prabhuji and Srivastava (1977a) observed entirely spherical or irregularly shaped resting sporangia, which were devoid of apical beak as that of *A. Javanicus* var. *allomorphus* (Isolate No. R.Q. 1/6 E). Misra (1982) found its occurrence in alkaline ponds of Lucknow.

Allomyces neo-moniliformis Indoh

Sci. Rept. Tokyo Bunrika Daigaku Sec. B **4** : 271; 1940

Syn. *A. cystogenus* var. *cystogenus* Emerson, Lloydia, **4** : 136, 1941

(Plate 2, Fig. 3)

Life cycle cystogenus type; plant large in size; zoosporangia 45-96.2 × 23-34.5 µm, secondary zoosporangia borne in long chains; encysted zoospores 10-12 μm in diameter; resting spores with rounded apex, 40-69 × 30-50.5 μm, averaging 54 μm in length and 40 μm in width, plants from resting spores 9-12 μm in diameter giving rise to 4 isogamous gametes, which fuse in pairs.

This species was isolated from soils of Ramgarh lake, Gorakhpur on hemp seed baits (Bhargava and Singh, 1965).

Allomyces recurvus Prabhuji and Sinha

Neo Botanica 1 (1 & 2) : 31-38, 1993

(Plate 2, Figs. 4-11)

Mycelium limited, dense; principal hyphae dichotomously branched, hyphal branches irregularly curved, 45-75 μm in diameter at base; zoosporangia ellipsoidal or ovoid, rarely spherical, terminal, sometimes catenulate, hyaline with apical and lateral discharge papillae; 60-120 μm long by 55-95 μm in diameter, renewed sympodially; resting spores abundant, ovoid, rarely spherical, terminal, thick walled, minutely punctate, yellow or brown in colour, 45-75 μm long by 37.5-46.5 μm in diameter; renewed sympodially, usually renewed hyphal branch coiled or recoiled round the resting spores before producing a new one; gametothalli (hyphae) similar to sporophytic hyphae, gametangial production moderately abundant, in terminal pairs; male gametangia lateral in position in relation to female gametangia, 30-45 μm long by 15.5-22.5 μm in diameter; pink to orange in colour, with apical discharge papilla; female gametangia broadly ellipsoidal, 57-85 μm long by 22.5-38.5 μm in diameter, hyaline with lateral discharge papilla (Prabhuji and Sinha, 1993).

It was collected from the soil samples of the environs of a water body near Chhawani area at Padrauna (District Deoria), U.P. in February 1991, leg, S.K. Prabhuji. The habitat was characterized by neutrally acidic soil (pH 6.5-7.0), moderate water content (27-30 per cent) and low organic contents (3-4 per cent), type culture in the Deptt. of Botany, St. Andrews College, Gorakhpur (WMSB-2206).

Plate – 1

Figs. 1-8 : *Allomyces anomalus* Emerson.

1. Thick, stout and branched basal cell attached to the substratum.

2-6. Fertile terminal branches showing globular zoosporangia.

7-8. Hyphae showing irregular surface.

Figs. 1-8, scale A.

Figs. 9-14 : *Allomyces arbuscula* Butler.

9. Branching pattern of hyphae.

10. Mature zoosporangium showing liberation of zoospores.

11. Thick-walled resting sporangium.

12 and 13. Terminal hyphae having gametangia (Dark coloured-male and light coloured female)

14. Empty gametangia.

Figs. 9, 10, 12 and 13, scale A; Figs. 11 & 14 Scale B.

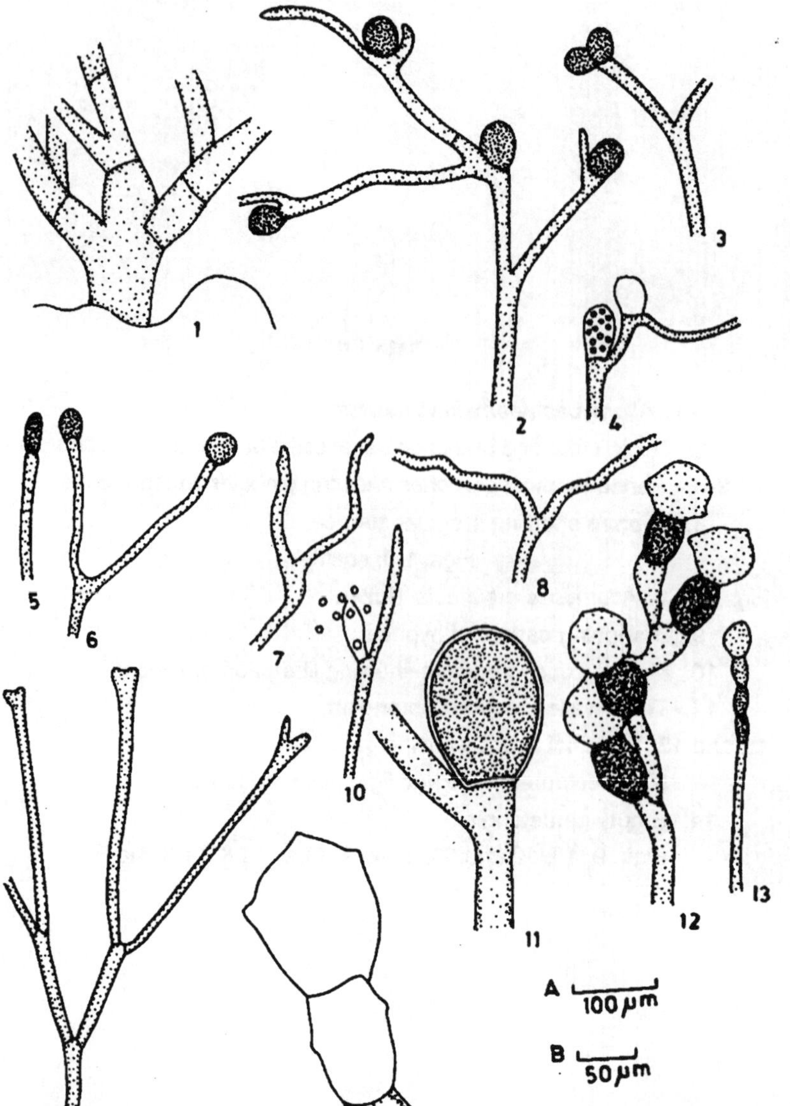

Plate 1

Plate-2

Fig. 1. *Allomyces javanicus* : Two apical sporangia in hyphae

Fig. 2. *A. moniliformis* : Basipetal secondary zoosporangia (Bhargava and Singh, 1965)

Fig. 3. *A. neo-moniliformis* : Chain of zoosporangia

Fig . 4–11- *A. recurvus*–Hyphae and sporangia (Prabhuji and Sinha, 1993).

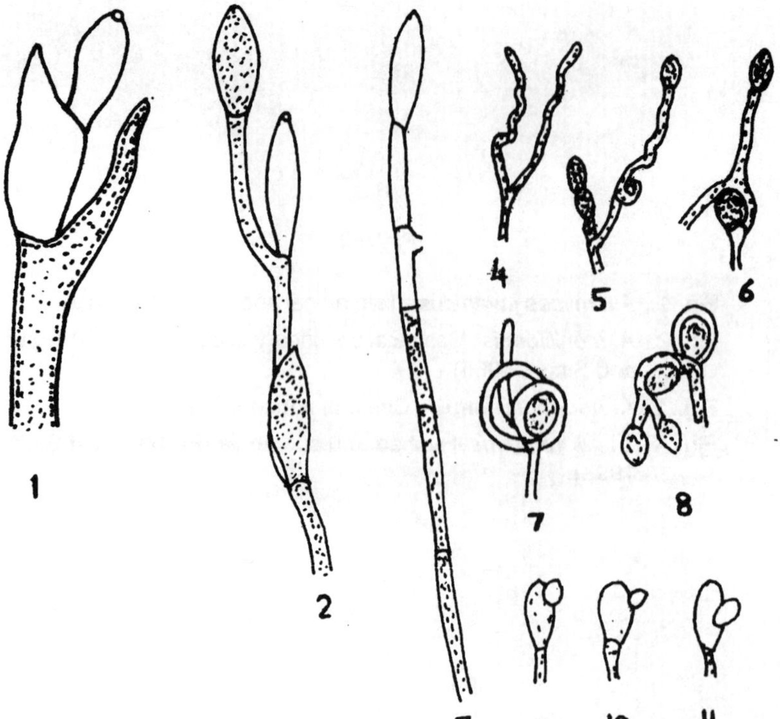

Plate 2

Blastocladia Reinsch

Jahrb. Wiss Bot., **11** : 298, 1878

Aquatic Phycomycetes, p. 678, 1960

Key to the Species of Blastocladia

Resting spores present .... A

Resting spores absent .... B

A_1 Habit arborescent C

A_2 Habit not arborescent.............................. *B. pusilla*, p. 27

B_1 Habit arborescent .... E

B_2 Habit not arborescent.............................. F

C_1 Basal cell stout and columnar .... *B. fruticosa*, p. 20

C_2 Basal cell not stout and columnar .... D

D_1 Basal cell slender and branched .... *B. mammillata*, p. 25

D_2 Basal cell long and cylindrical .... *B. arborata*, p. 15

E_1 Basal cell columnar and trunk like...... *B. sessilis*, p. 29

E_2 Basal cell not so .... R

F_1 Basal cell variable, swollen lobes as distal part................................ B. *pringsheimii*, p.27

F_2 Basal cell not variable, no swollen lobes.. G

G_1 Branches glomerate.......................... *B. glomerata*, p.23

G_2 Branches not glomerate.................... H

H_1 Basal cell globose ....*B. globosa*, p. 22

H_2 Body ramose.. I

H_3 Body not ramose.................................. J

I_1 Two types of sporangia (small and large).. K

I_2 Sporangia one type .... *B. incrassata*, p.24

J_1 Basal cell columnar, trunk like........... *B. spicaformis*, p.30

J_2 Basal cell neither columnar nor trunk like.. M

K_1 Sporangia hyaline .... *B. excelsu*, p. 19

K_2 Sporangia not hyaline L

L_1 Large sporangia rare, fewer, filled with contents .... *B. coronata*, p. 17

L_2 Large sporangia empty, common .... *B. sparrowii*, p. 30

M_1 Sporangia on compound spike .... *B. elegans*, p. 18

M_2 Sporangia on simple branches .... N

N_1 Sporangia bilobed .... *B. didyma*, p. 17

N_2 Sporangia not bilobed .... O

O_1 Sporangia ten times longer than breadth...P

O_2 Sporangia smaller .... Q

P_1 Sporangia curved....................................... *B. filamentosa*, p.19

P_2 Sporangia not curved................................ *B. angusta*, p.15

R_1 Basal cell slender and filamentous .... S

R_2 Basal cell cylindrical .... T

S_1 Sporangia one type, brown .... *B. fusiformis*,p. 21

S_2 Sporangia two types .... *B. heterosporangia*, p. 23

T_1 Sporangia two types, larger sporangia deciduous .... *B. caduca*, p. 16

T_2 Sporangia one type only .... U

U_1 Sporangia on short protuberances of branches, rostrate....................................... *B. rostrata*, p. 29

U_2 Sporangia mostly terminal, some are subterminal.. V

V_1 Empty sporangia persistent .... *B. ramosa*, p. 28

V_2 Sporangia pitcher shaped, not persistent.. *B. pileota*, p. 26

1. *Blastocladia angusta* Lund

Kgl. Danske, Seink. Skrift, Naturv. Math., Afd., **6** : 44, 1934.

(Plate 3; Fig. 25)

Thallus cylindrical, branched, 150-500 μm; proximal part swollen, bearing cylindrical, long (250 × 25 – 35 μm), sessile, smooth walled, hyaline zoosporangia; basal cells cylindrical, 100–120 × 20 – 30 μm; rhizoids hairy, branched, setae present or absent, long, black, unbranched; zoospores liberation and resting body formation not observed (Misra and Dwivedi, 1987).

It was isolated from pericarps of guava fruits submerged in lake water (Ingold No. AP 26), Varanasi (Mishra and Dwivedi, 1987). Thalli of this isolate were mixed with the population of *B. sparrowii* and *B. ramosa*. Here, sporangia were about 9-10 times longer than the width and produced at the top of the crowns of the thalli.

2. *Blastocladia arborata* Das Gupta and John

Indian Phytopathology **41** (4) : 521-547, 1988

(Plate 3; Figs. 1-3)

Plant arboraceous, variable in size and shape, whole thallus up to 400 μm high, the largest 450 μm, wide at the crown; basal cell long, cylindrical, smooth walled, 100-140 × 35 μm in diam. with occasionally swollen base or stunted proximatly, 85-90 μm in diam; the distal end dividing dichotomously usually forming two divergent cylindrical appearing as cleft thallus or forming more axes, which further divide cymosely forming successively slender branches at the top; holdfasts arise from the base, variable, moderately stout or delicate, sparsely branched; sporangia borne terminally on small protuberances or at the tip of the slender branches, oblong or fusoid, occasionally basal end constricted, base truncate, apex convex, 20-45 × 12-18 μm in diam. where widest, discharge pore formed at the tip, contents highly refractive, developing into large number of zoospores at maturity, zoospores spherical; flagellum posterior, emerge singly through the pore and swim away; resting spore borne terminally or subterminally, singly alongwith sporangia, spherical, 20-25 μm in diam; base truncate, double walled, outer wall thick, punctate, inner wall thin, faintly brown, germination not observed (Das Gupta and John, 1988).

Saprophytic, appearing as small pustules on the styles of the submerged flowers of *Caryopteris* sp. as well as in Myrtaceous fruit used as bait in the tanks in the garden of the Botany Department, Lucknow University, Lucknow. Holotype Camera lucida drawings deposited with the Herbarium, Division of Mycology and Plant Pathology, IARI, New Delhi 110 012. HCIO No 40, 581 (Figs. 1-3) (Das Gupta and John, 1988).

3. *Blastocladia caduca* Das Gupta and John

Indian Phytopathology **41**(4): 521 - 547, 1988

(Plate 4; Figs. 36 - 38)

Habit tree-like, whole thallus up to 350 μm high; basal cell uniformly cylindrical, 13-140 × 30-35 μm in diam., distal end gives rise to alternate lateral axes and several upright axes at its apical region, the dichotomously developed axes cylindrical, 230 × 30 μm wide, unbranched, smooth walled for a long distance, very spare, dividing terminally into slender branches, rarely lobes holdfasts moderately stout, branches arising from the base; sporangia of two types, those smaller terminal, borne single on the tips of branches or lobes, ovoid or narrow fusiform, 35-40 × 12-15 μm in diam. hyaline, sessile base truncate, apex convex, prominent apical pore, deciduous, the second type larger sub-terminally borne on the tip of deciduous, the second type larger subterminally borne on the tip of lateral protuberance (suppressed axis), 60 -70 × 15-20 μm in diam., hyaline, sessile, with truncate base and prominent apical pore; zoospores sub-globose, posteriorly uniflagellate, 5.2 μm diam. in smaller sporangia, 7-8 μm in larger ones; resting spore not found (Das Gupta and John, 1988).

Saprophytic appearing as small whitish pustules on the surface of the stamens of the submerged flowers of *Crataeva religiosa* used as baits in the tanks in the garden of the Botany Department, Lucknow, University, Lucknow. Holotype deposited as Camera lucida drawings with the Herbarium, Division of Mycology and Plant Pathology, IARI, New Delhi-110 012. HCIO No. 40, 596 (Das Gupta and John, 1988).

4. *Blastocladia coronata* Das Gupta and John

Indian Phytopathology **41**(4) : 521 - 547, 1988

(Plate 4; Figs. 39-43)

Plant ramose, tall, whole thallus measuring up to 300 μm; basal cell narrow, cylindrical, 60-65 × 12-15 μm in diam., base giving rise to moderately thick, branched holdfasts, proximal end divides dichotomously to give rise to several upright, main axes, cylindrical, 10-12 μm in diam., each remains undivided for a considerable upright, narrow cylindrical, gradually widening to form an expanded globose head, occasionally apical and divides dichotomously to form short branches with swollen head, on which sporangia are borne; sporangia of two types, those borne on the expanded to the plant, or in irregular series, small, ovoid, 15-18 × 6-8 μm in diam. with prominent pore at the tip; the second type larger, rare, fewer in number, borne singly at the axil of the branches, pitcher shaped (ascidium), curved, attenuated towards the apex or irregularly cylindrical, 30-35 × 10-12 μm in diam., base attenuated or rounded, apical pore prominent; zoospores and their mode of discharge not studied; resting spore not observed (Das Gupta and John, 1988).

Saprophytic appearing as pustules on the surface of the fruit of *Eugenia jambolana* used as bait in the tank in the garden of the Botany Department, Lucknow University, Lucknow (Das Gupta and John, 1988).

5. *Blastocladia didyma* Das Gupta and John

Indian Phytopathology, **41**(4) : 521-547, 1988

(Plate 3; Fig. 29)

The whole thallus 450 μm high; basal cell wide, cylindrical, smooth walled, 250 × 90 μm wide, divides distally into two main axes dichotomously, 30 μm wide, each one further divides also dichotomously once or twice, also sympodially in some with branches on the either side suppressed, ultimate branches 20-25 μm; holdfasts basal, copious, much branched, slender, penetrating deep into substratum; sporangia arise terminally directly from ultimate branches, continuous with it, without the formation of cross septum delimitting the reproductive body, borne as a Siamese twin, fused or undivided up to different levels from the base onwards to the tip, exhibiting bisected, bipartite or bilobed condition, with discrete lobe widely apart, occasionally fully fused right up to the tip, each lobe or the sporangium long, cylindrical, ovoid, allantoid, 70-180 × 20-30 µm wide; when fully fused long, fusiform, 130-180 × 30-35 µm wide, a second type of sporangium much rarer, long, pyriform with elongated tip, 130-150 × 40-50 µm wide also borne on terminal axes, delimited by a septum; zoospores and their mode of discharge in either type of sporangia not observed; resting spore not observed (Das Gupta and John, 1988).

Saprophytic on the surface of the submerged fruit or Tipari (*Physalis peruviana*) used as a bait in the tanks of the garden of the Botany Department, Lucknow University, Lucknow. Holotype deposited as Camera lucida drawings with the Herbarium, Division of Mycology and Plant Pathology, IARI. New Delhi 110012. HCIO No. 40,592. (Fig. 29) (Das Gupta and John, 1988).

6. *Blastocladia elegans* Das Gupta and John

Indian Phytopathology, 41(4) : 521–547, 1988
(Plate 3; Figs. 21 and 22)

Plant slender, elegant; whole thallus 250 µm in length, basal cell 100-110 × 15–20 µm in diam., uniformly cylindrical, slightly expanded base gives rise to slender, branched holdfasts, axis arising from distal end of the basal cell a straightened scorpioid sympodium, formed by more or less complete suppression of successive axes on alternate sides at successively shorter intervals in dichasium form of branching, sporangia borne on slender, diminutive branch of suppressed axes along the sympodium somewhat in the form of a compound spike, ovoid or fusiform, 22–28 × 9-11 µm in diam., sessile, hyaline, deciduous, base attenuated, truncate apex convex, discharge pore formed by the dissoulution of apical end; zoospores and their mode of discharge not observed, resting spore not found (Das Gupta and John, 1988).

Saprophytic on the surface of the submerged mango fruit baited in the tanks of the garden of the Botany Department, Lucknow University, Lucknow. Holotype deposited as Camera lucida drawings with the Herbarium, Division of Mycology and Plant Pathology, IARI. New Delhi–110012. HCIO No. 40, 584. (Fig. 29) (Das Gupta and John, 1988).

7. *Blastocladia excelsa* Das Gupta and John

Indian Phytopathology, **41**(4) : 521-547, 1988

(Plate 4; Figs. 33-35)

Whole thallus up to 250 μm high, basal cell 120 μm, cylindrical, holdfasts arising at the base, moderately thick, branched, distal end divides into two or more axes, cylindrical, relatively short, branching dichasial cyme, axes at short intervals suppressed, forming successive small protuberances on which sporangia are borne; sporangia of two types, smaller narrowly or broadly fusiform, 30-37 μm × 10-13 μm wide, sessile, hyaline, apex constricted, pore apical, 4-5 μm in diam.; persistent, retaining normal shape when empty, the larger type cylindrical, hyaline, 50-54 × 8-13 μm wide, sessile with constricted apex and prominent wide pore or clavate, 65 × 10 μm proximally where widest, tapering distally, apex convex, refractive contents divide to form zoospore initials which mature within the sporangium and emerge singly through the pore; zoospore spherical, posteriorly flagellate as observed in smaller type, while zoospores and their mode of discharge in larger type not observed; resting spore not found (Das Gupta and John, 1988).

Saprophytic on the surface of the ripe guava fruit used as bait in the tanks in the garden of the Botany Department, Lucknow University, Lucknow. Holotype deposited as Camera lucida drawings with the Herbarium, Division of Mycology and Plant Pathology, IARI, New Delhi-110012 HCIO No. 40, 597. (Figs. 33-35) (Das Gupta and John, 1988).

8. *Blastocladia filamentosa Das Gupta and John*

Indian Phytopathology, **41**(4) : 521-547, 1988

(Plate 3; Figs. 27 and 28)

Whole thallus up to 400 μm high; basal cell variable in shape and size, cylindrical, 150 × 30 μm broad or bulbous, 100 × 70 μm broad, distally divided dichotomously into two main axes, each of which divides once, rarely twice, also dichotomously forming longer branches, cylindrical, 100 × 20 µm wide, the apical end bifurcates into two lobes, swollen, or branch as dichasial cyme with suppressed axes at very short intervals; holdfasts thin, filamentous all over the base, branched; sporangia borne singly on the top of the lobes or on the suppressed axes, in the latter condition appearing as cluster, narrow, 100-140 × 10-15 µm in diam., 10 times as long as broad or longer, slightly curved, sessile, hyaline, base truncate, apex convex, that occasionally buds, proliferate apically to form 2-3 appurtenances terminating in papillulate structures, or give rise to several short excrescences in parallel series, alongwith these, but in separate branch, rarely a second type of sporangia occur, cylindrical or allantoid, 40-80 × 10-12 µm wide, some of which also proliferate; zoospores and their mode of discharge, the discharge pore were not observed, resting spore not found. (Das Gupta and John, 1988).

Saprophytic appearing as small pustules on the surface of *Eugenia jambolana* fruit, used as bait in the tanks in the garden of the Botany Department, Lucknow University, Lucknow. Holotype deposited as Camera lucida drawings with the Herbarium, Division of Mycology and Plant Pathology, IARI, New Delhi 110012. HCIO No. 40, 594 (Figs. 27, 28) (Das Gupta and John, 1988).

9. *Blastocladia fruticosa* Das Gupta and John

Indian Phytopathology, 41(4) : 521-547, 1988

(Plate 3; Figs. 4 and 5)

Habit arboraceous; whole thallus 150 × 250 µm in diam., crown semi circular; basal cell 30 × 50 µm wide, stout, columnar, dwarfish, thick walled, base divides into thick branched holdfasts, distal end forms the main axis, 60 × 75 µm high giving rise to alternate, lateral and vertical branches, which divide into two or more thick lobes, sporangia borne terminally and subterminally on successive lateral, closely spaced protuberances or lobes, mainly pyriform, 28-30 × 15-17 µm in diam., or cylindrical, 25-30 µm long by 9-10 µm wide, sessile, apex broadly convex, deciduous; zoospores and their mode of discharge not observed;

resting spores arise intermingled with sporangia successive ones forming long chain, spherical, 15-25 μm in diameter, rarely ovoid, 20-22 × 17 μm in diameter, slightly attenuated at the base, double walled, inner thin, smooth, outer thick, pitted, brown in colour, in mature when sporangia are shed off the chains of persistent resting spores form a conglomeration of clusters covering the entire branching system; germination of resting spore not observed (Das Gupta and John, 1988).

Saprophytic appearing as small whitish pustules on the surface of the submerged fruits of *Eugenia jambolana*, used as bait in the tanks in the garden of the Botany Department, Lucknow University, Lucknow. Holotype as Camera lucida drawings deposited with the Herbarium, Division of Mycology and Plant Pathology, IARI, New Delhi-110012. HCIO No. 40, 582. (Figs. 4.5) (Das Gupta and John, 1988).

10. *Blastocladia fusiformis* Das Gupta and John

Indian Phytopathology, **41**(4) : 521-547, 1988

(Plate 4; Figs. 30-32)

Plant tree-like, delicate, whole thallus up to 250 μm high; basal cell 150 μm long, expanded at the base (40 μm), gradually tapering upwards, 20 μm wide at the top, distal end expands and divides dichotomously giving rise to two main axes, rarely undergoing further divisions into smaller axes, cylindrical, wall smooth, unbranched, tip bifurcates often repeatedly forming several short, slender branches or lobes; pseudosepta present in one of the main axes; holdfasts long, thin, slender, branched; sporangia borne singly on lobes or on ultimate slender branches, forming a cluster, broadly fusiform, 35-40 × 15-18 μm in diam., brownish, pore apical, base truncate; zoospores and their mode of discharge not observed; resting spore not found (Das Gupta and John, 1988).

Saprophytic, appearing as small whitish pustules on the surface of the submerged stamens and pistils of the flower of *Crataeva religiosa* used as baits in the tanks in the garden of the Botany Department, Lucknow University, Lucknow. Holotype deposited as Camera lucida drawings with the Herbarium, Division of Mycology and Plant Pathology, IARI, New Delhi-

110012. HCIO No. 40, 589, (Figs. 30-32) (Das Gupta and John, 1988).

11. *Blastocladia globosa* Kanouse

Amer., J. Bot. **14** : 298, 1927

(Plate 5; Fig. B)

Basal cell globose, proximal cylindrical part very short, 70-400 µm long by 70-300 µm broad, distantly entire or with large irregular expanded lobe, 100-70 µm wide, wall smooth thick, 8-12 µm, contents brownish; holdfast stout, branched, setae absent, whole plant 260 µm long ; sporangia broadly clavate, 90-190 µm long by 32-40 µm broad, numerous; resting spores not observed (Das Gupta and John, 1953).

This species was isolated on submerged fruits from pond water, Lucknow (Das Gupta and John, 1953); Ramgarh lake and University garden pond, Gorakhpur (Bhargava and Singh 1965); Brahma Sarovar, Kurukshetra (Gupta and Mehrotra, 1989a, b, 1992); from alkaline pond water, Lucknow (Misra, 1982), and on submerged guava, rose apple from lake water, Varanasi (Misra and Dwivedi, 1987). Misra and Dwivedi (1987) found more than one globose heads on the basal cells, many dark colored setae on the basal heads of the thalli.

12. *Blastocladia globosa* var. *bhargavi*

This species was isolated on mango fruit from pond A, Hyderabad during winter (October to January) (Manoharachary and Rao, 1981).

13. *Blastocladia globosa* var. *minutissima* Manoharachary, Bhairabnath and Rao

Bibliotheca Mycologica **91** : 460-641, 1983

Thallus 320-650 µm long, basal cell globose, 36-126 µm in diam., the proximal part short, smooth; holdfast stout; setae absent; zoosporangia cylindrical, 21-54 × 11-36 µm; zoospores spherical, posteriorly uniflagillate; sexual state not observed (Manoharachary *et al.*, 1983).

This species formed minute white pustules on unopened mango fruit bait from temple pond water, Vikarabad, Hyderabad, A.P. July 1978, at pH 6.2, water temperature 24.0°C, OUBF B (Manoharachary *et al.*, 1983).

14. *Blastocladia glomerata* Sparrow

J. Linn. Soc. London (Bot.), **50**: 456, 1936

(Plate 3; Figs. 18 and 19)

Plant stout; whole thallus measuring up to 300 µm high; basal cell thick, columnar, wide at the base, tapering upwards, 200 × 60 µm wide, 80 µm widest, or cylindrical throughout, distally divides to form two thick stout lobes, 90 × 60 µm wide, each bifurcating into two swollen lobes, or form two cylindrical branches, each ending in one or more swollen lobes, holdfasts arising from basal end, moderately stout, branched; sporangia ellipsoidal, ovoid or narrow, fusiform, 70-93 × 22-50 µm in diam., length and breadth ratio 1: 2.5-3, hyaline, sessile, attenuated at the proximal end, borne in clusters on the surface of the swollen lobes, attached by the truncate base, old sporangia deciduous, leaving scar on the surface when shed; setae absent; zoospores and their mode of discharge not observed; resting spore not found (Das Gupta and John, 1988).

Saprophytic appearing as small whitish pustules on the surface of the baited fruits of *Eugenia jambolana*, in the tanks of the garden of the Botany Department, Lucknow University, Lucknow. Holotype deposited as Camera lucida drawings with the Herbarium, Division of Mycology and Plant Pathology, IARI, New Delhi-110012. HCIO No. 40, 590. (Figs. 18-19) (Das Gupta and John, 1988).

15. *Blastocladia heterosporangia* Das Gupta and John

Indian Phytopathology, **41**(4) : 521-547, 1988

(Plate 4; Figs. 44 and 45)

Plant arboraceous, whole thallus up to 400 µm high; basal cell wide, stout, columnar, 100-125 × 45-75 µm broad, greatest width at the base, tapering upwards, 50 µm wide at the distal end, expands and divides dichotomously into several axes, 130-150 × 20-30 µm in diam., long, smooth walled, unbranched, apex bilobed or gives rise to short slender axes with bilobed swollen tips; pseudosepta present but rare; holdfasts arising from basal end, filamentous, branched; sporangia of two types, those long cylindrical, tapering distally, 50-90 × 10-12 µm in diam., sessile, hyaline, numerous, apex convex, base truncate, borne terminally on the swollen lobes, the second type smaller, fewer in number, borne also terminally from the adjoining base as the larger sporangia, alongwith them, ovoid or oblong or ascidial in shape, persistent empty sporangia retain normal ovoid shape, 25-30 × 12-15 µm in diam., apex convex, base rounded, truncate; zoospores and their mode of discharge in cylindrical ones not observed; in smaller, ovoid or oblong type zoospores emerge in mass, further stages of their development not observed; resting spore not found (Das Gupta and John, 1988).

Saprophytic, appearing as small white pustules on the surface of the sumerged grape fruit baited in the tranquil water of the river Gomti flowing by the side of the Lucknow University, Lucknow (Das Gupta and John, 1988).

16. *Blastocladia incrassata (Indoh)* Das Gupta and John

Science Report, Tokyo Bunrika Daigu, Sect. B, **4** : 232, 1940.
(Plate 3; Fig. 20)

Plant densely ramose; whole thallus up to 500 µm in length; basal body cylindrical, smooth, thin walled, 350 × 50 µm in diam., distal end divides dichotomously into 3-4 main axes, each 30-40 µm wide, which branch further, giving rise to cylindrical branches or form thick terminal lobes, further branching sympodial cyme; holdfasts arising from the basal end, copious, slender, much branched; sporangia terminal, formed singly on the tip of short branches, allantoid, cylindrical, one side incurved, 40-92 × 20-28 µm in diam., base wide, rounded, apex convex, or long, pyriform constricted at the neck, 90-100 × 20-30 µm in diam. where widest, rarely 130 × 20 µm wide; zoospores and their mode of discharge not observed; resting spore not found (Das Gupta and John, 1988).

Saprophytic on the submerged fruits of Guava used as a bait in the tanks in the garden of the Botany Department, Lucknow University, Lucknow. Camera lucida drawings deposited with the Herbarium, Division of Mycology and Plant Pathology, IARI, New Delhi-110012. HCIO No. 40, 591. (Fig. 20) (Das Gupta and John, 1988).

17. *Blastocladia mammillata* Das Gupta and John

Indian Phytopathology, **41**(4) : 521-547, 1988

(Plate 3; Figs., 15-17)

Plant tree like, whole thallus up to 400 µm high; basal cell cylindrical, 80-110 × 25-30 µm in width, somewhat conical, tapering from base to the tip, distally divides forming 3-4 upright axes, which give rise to several lateral branches, 60-85 × 20 µm diam., unbranched for considerable length, again dividing dichotomously in subterminal region once or twice to form slender terminal branches; pseudosepta present; holdfasts arise at the base of the basal cell, slender, branched; sporangia terminal, borne singly on the tip of slender branches, irregularly alternate, or on successive protuberances (suppressed axes), at short intervals, on the same or both sides forming a cluster, developing in an irregular series forming a crown of irregular mass, broadly spindle shaped, 23-30 × 18-23 µm in diam., sessile, hyaline, truncate at the base, apex with prominent papilla, pore apical; zoospores and their mode of discharge not observed; resting spores formed alongwith the sporangia apically on the tip of terminal slender branch ellipsoidal, 20-35 µm long × 13-15 µm in diam., sessile, brownish, wall moderately thick, minutely pitted; germination of resting spores not observed (Das Gupta and John, 1988).

Saprophytic, appearing as small pustules on the surface of the submerged *Pyrus* fruit, used as bait in the tanks in the garden of the Botany Department, Lucknow University, Lucknow. Holotype deposited as Camera lucida drawings with the Herbarium, Division of Mycology and Plant Pathology, IARI, New Delhi-110012. HCIO No. 40, 586. (Figs. 15-17) (Das Gupta and John, 1988).

18. *Blastocladia pileota* Das Gupta and John

Indian Phytopathology, **41**(4) : 521-547, 1988

(Plate 3; Figs. 9-11)

Plant arboraceous, whole thallus up to 460 µm, basal cell more or less uniformly cylindrical, 160-230 × 50-60 µm in diam., branching dichotomously at the distal end giving rise to three or four cylindrical axes, somewhat swollen at the base, which further divides dichotomously once or twice into smaller slender branches; holdfasts formed at the base, filamentous, branched; sporangia terminal borne singly or in twos or threes on the tips of the protuberances formed on the club shaped terminal branch, sessile, broadly ovoid, pitcher shaped, 40-68 × 20-30 µm where widest, broad and rounded at the base, narrowing gradually towards the apex, constricted at the subapical region, 5 µm wide, swollen apex, surrounded by an ellipsoidal, cap-like papilla, 6 × 10 µm wide; highly refractory, heterogenous contents retract from wall before discharge, when empty wall assumes narrow, ovoid or fusiform with wide open apex, the mode of formation of pore whether by the dissolution of the apical cap or by its ejection after rupture or circular rent not observed: zoospores, their formation, their maturity and mode of discharge not observed; resting spore not found (Das Gupta and John, 1988).

Saprophytic, appearing as small whitish pustules on the surface of the submerged fruit of *Pyrus* used as bait in the tanks in the garden of the Botany Department, Lucknow University, Lucknow. Holotype deposited as Camera lucida drawings with the Herbarium, Division of Mycology and Plant Pathology, IARI, New Delhi-110012. HCIO No. 40, 587. (Figs. 9-11) (Das Gupta and John, 1988).

19. *Blastocladia pringsheimii* Reinsch

Jahrb, wiss Bot. **11** : 298, 1878

(Plate 5, Fig. 1; Plate 6, Figs. 1-2)

Thallus 570-715 µm, stout, basal cell variable in size, 250-420 × 70-105 µm, generally cylindrical with swollen lobes on its distal part; zoosporangia 70-150 × 15-30 µm, not proliferating, empty sporangia remain intact for a long time; zoospores spherical or ovoid, 7-13.8 µm in diam., long setae present. Kiran and Dayal (1992) isolated *Blastocladia pringsheimii* on mango fruit from the small ponds of Agricultural farms of the Institute at B.H.U. Their isolate was characterized by stout unbranched thallus, absence of setae, sporangia on the tip of the branches or over the surface, cylindrical, 70-350 µm long and 13-60 µm in diameter, discharge papilla with endogenous plug, abundant resting spores, ovoidal or spherical with truncate base, 30-50 µm in diameter, thick walled brownish and punctate (Bhargava and Singh, 1965).

It was isolated from apple, grape, cape, gooseberry and tomato baits, Rapti River, Ramgarh lake and University garden pond, Gorakhpur (Bhargava and Singh, 1965); water samples of Hanuman Tal pond, Jabalpur during winter, summer, monsoon and post monsoon (Agarwal *et al.*, 1989) and from decaying plant material and bottom sediments of polluted waters of Ganga Sagar, Jabalpur (Mishra *et al.*, 1990).

20. *Blastocladia pusilla* Das Gupta and John

Indian Phytopathology **41**(4) : 521-547, 1988

(Plate 3; Figs. 6-8)

Whole thallus up to 250 µm high; a small, simple, slender plant; basal cell cylindrical, 50-60 × 15-18 µm in diam., dividing dichotomously at successive intervals forming several axes, erect, compact, cylindrical, smooth walled, apical branching dichasial cyme; holdfasts arising from the end of the basal cell, delicately branched; sporangia borne on the tip of the protuberances (suppressed axes), terminal, sessile, broadly spindle shaped, 25-30 × 15-18 µm in diam. at the middle where widest, 5.8 µm in diam. at the base and the apex, base truncate, apex with knob-like papilla opening by a pore, occasionally a peg-like plug occludes the pore and projects inside the sporangium; resting spore terminal, similarly borne as the sporangia, mingled with them, subspherical, 17-20 × 15-18 µm in diam. with rounded apex, basal end slightly attenuated, base truncate, sessile, brown, inner wall smooth, thin, outer wall thick, punctate: germination of resting spore not observed (Das Gupta and John, 1988).

Saprophytic, developing as small whitish pustules on the surface of the various submerged fruits of jamun & guava, used as baits in the tank of the garden of the Botany Department, Lucknow University, Lucknow. Holotype deposited as Camera lucida drawings with the Herbarium, Division of Mycology and Plant Pathology, IARI, New Delhi-110012. HCIO No. 40, 585. (Figs. 6-8) (Das Gupta and John, 1988).

21. *Blastocladia ramosa* Thaxter

Bot. Gaz. **21** : 50, 1896

(Plate 5, Figs. C & D)

Habit tree like, basal cell cylindrical, 250-600 µm long by 90 µm in diam., divided distantly, dichotomously into several secondary axes which branch and rebranch profusely; holdfast delicate, sparingly branched, wall thin, smooth, setae absent, whole thallus with an open ramose habit up to 570 µm long; sporangia terminal and subterminal on short branches appearing like ear of wheat, long cylindrical, empty sporangia persisting, 32-45 µm long by 10-16 µm broad, papilla apical with an apical pore; zoospores spherical; resting spores generally not observed (Das Gupta and John, 1953).

Isolated on submerged fruit from Lucknow (Das Gupta and John 1953) and on apple and grape fruit from Rapti river, Ramgarh lake, University garden pond water of Gorakhpur (Bhargava and Singh, 1965). Manoharachary *et al.* (1983) isolated it from water of Alisagar lake, Nizamabad (pH 8.0, temperature 26°C) on mango fruit bait , May 1978 OUBF 126/A. This isolate showed longer and wider zoosporangia (24-60 µm long, base 8-16 µm wide; middle portion 12.8-19.2 µm wide; apical 4.8-12.8 µm wide and resting sporangia (16-36.8 µm long, base 8-11. 2 µm wide; middle portion 22.4-24 µm wide; apical 12.8-24 µm wide).

Misra (1982) isolated it from alkaline ponds of Lucknow. This species was isolated on submerged guava and rose apple fruit in lake water, Varanasi (Misra and Dwivedi, 1987). They observed resting sporangia like zoosporangia, which were spherical, broadly ovoid or clavate with round apex and narrow base, 25-40 µm, smooth walled, wall slightly thick and constricted. They did not observe germination of resting spores.

22. *Blastocladia rostrata* Minden

Proc. of the Indian Acad. Sciences **38**(4) B : 165-170, 1953

(Plate 5, Fig. E)

Habit slender, long tree like, basal cell narrowly cylindrical, 200 µm long by 32 µm in diameter, distantly branching dichotomously, branches cylindrical throughout; holdfast quickly branched, setae absent, whole plant 850 µm long; sporangia alternately arranged along the branches, on short protuberances, the younger ones being at the top, empty sporangia persistant long and cylindrical with an opening at the tip; 60-70 µm long by 12-20 µm broad, young sporangia club shaped up to 40 µm long; resting spores not observed (Das Gupta and John, 1953).

Isolated on submerged fruits (Das Gupta and John, 1953).

23. *Blastocladia sessilis* Das Gupta and John

Indian Phytopathology **41**(4) : 521-547, 1988

(Plate 3; Figs., 23 and 24)

Habit arboraceous, whole thallus up to 200 µm high, widely spreading; basal cell wide, short, columnar, trunk-like, 60-80 × 40-45 µm in diam., distal end divides dichotomously to give rise to several cylindrical, somewhat scorpioid axes, subsequent branching dichasial cyme, terminal end lobed; sporangia borne terminally, one on each lobe or along the fertile axes on alternate, lateral suppressed axis, directly on it, singly, broadly ovoid, 20-22 × 10-12 µm in diam., sessile, base truncate, apex convex, discharge pore prominent, 4 µm wide, persistent even when empty; zoospores ellipsoidal, posteriorly uniflagellate; resting spore not observed (Das Gupta and John, 1988).

Saprophytic, developing as small whitish pustules on the surface of the various submerged fruits, jamun and guava used as baits in the tanks of the garden of the Botany Department, Lucknow University, Lucknow. Holotype deposited as Camera lucida drawings with the Herbarium, Division of Mycology and Plant Pathology, IARI, New Delhi, 110012, HCIO No. 40, 585, (Figs. 23-24) (Das Gupta and John, 1988).

24. *Blastocladia sparrowii* Indoh

Science Rept. Tokyo Bunrika Daigaku, Sec. B., **4**:259.1940.
(Plate 5; Fig. F)

Basal cell cylindrical, 200 µm long by 8 µm in diam.; wall smooth, expanded, distantly giving rise to cylindrical subdichotomously arranged lobes; setae absent; holdfast branched; the whole plant 450-500 µm long; sporangia borne over the surface of the lobes of two types, bigger empty ones and smaller ones filled with content, the smaller ones cylindrical, tips narrow, curved, 60-140 µm long by 20-40 µm in diam., bigger ones cylindrical, the narrow tips having prominent apical papilae,110-140 µm long by 23-30 µm broad; resting spores generally not observed (Das Gupta and John, 1953).

Isolated on submerged fruit from Lucknow (Das Gupta and John, 1953), from Brahma Sarovar, Kurukshetra (Gupta and Mehrotra, 1989 a, b, 1992). Misra and Dwivedi (1987) isolated it on pericarps of guava fruits from fresh water, Varanasi. They observed abnormal swelling at the distal part of basal cells, which were not found in any other Indian isolate from fresh water non–alkaline habitats. They also reported presence of rare resting spores and sterile, unbranched, septate setae (100-250 µm long) in their isolate.

25. *Blastocladia spicaformis* Das Gupta and John

Indian Phytopathology **41**(4) : 521-547, 1988
(Plate 3; Figs. 12-14)

Whole thallus 525 µm high; basal cell columnar, trunk-like, 250-300 × 90-120 µm broad, basal end 85 µm wide, bifurcate into two lobes of unequal width, distal end 110 µm wide, divides dichotomously into 2-3 axes, about 60 µm wide, each giving rise to several cylindrical axes, 170-200 × 15-25 µm in diam., branching dichasial cyme, forming erect scorpioid sympodia, fascicles of compound spike like form of axes rendering thallus its characteristic appearance; holdfasts copious, moderately slender, branched; pseudosepta present in certain branches; sporangia terminal and subterminal, borne singly on the protuberance of the bifurcated or on the tip of the alternating suppressed axis, narrowly fusiform, 25-35 × 7.11 µm in diam., base truncate, apex rounded, tip evanesces to form discharge pore; zoospores and their mode of discharge not observed; resting spore not observed. (Das Gupta and John, 1988).

Saprophytic, appearing as a small pustule on the surface of the Guava fruit used as baits in the tanks of the garden of the Botany Department, Lucknow University, Lucknow. Holotype deposited as Camera lucida drawings with the Herbarium, Division of Mycology and Plant Pathology, IARI, New Delhi-110012, HCIO No. 40, 588, (Figs. 12-14) (Das Gupta and John, 1988).

Plate - 3

(Das Gupta and John, 1988).

Figs. 1-3,	*Blastocladia arborata*: 1- habit (X140); 2- sporangia (X900), 3- resting spores with cluster of sporangia (X900).
Figs. 4-5,	*B. fruticosa*: 4- habit (X200), 5-sporangia and resting spores (X900).
Figs. 6-8,	*B. pusilla*: 6- habit (X200), 7-8- sporangium, apical pore with a plug and resting spores (X900).
Figs. 9-11,	*B. pileota* : 9-habit-highly infected (X 240), 10-sporangia (X600), 11-sporangia highly magnified (X900).
Figs. 12-14,	*B. spicaformis* : 12-habit showing spoke like branches and psuedosepta (X200). 13 and 14- sporangia (X900).
Figs. 15-17,	*B. mammillata* : 15-habit (X240), 16-sporangia (X600), 17-sporangia with resting spores (X 600).
Figs. 18-19,	*B. glomerata* : 18- habit (X 200), 19- sporangium (X 900).
Fig. 20,	*B. incrassata* : 20-habit (X 200).
Figs. 21-22,	*B. elegans* : 21- habit (X 200), 22-sporangia (X 900).
Figs. 23-24,	*B. sessilis* : 23-habit (X 200), 24-sporangia alternately and terminately borne (X 900).
Figs. 25-26,	*B. angusta* : 25-habit (X 200), 26-sporangia showing basal attachment (X 900).
Figs. 27-28,	*B. filamentosa* : 27 and 28-habit showing sporangia (X 140).
Fig. 29,	*B. didyma* : 29-habit with sporangia (X 140).

Plate 3

Plate - 4

(Das Gupta and John, 1988)

Figs. 30-32, *Blastocladia fusiformis* : 30-habit showing pseudosepta (X 140), 31-sporangia (X 900), 32-sporangia (X 240).

Figs. 33-35, *B. excelsa* : 33-habit (basal end not shown) with sporangia (X 600), 34-a branch with smaller sporangia (X600); 35-showing longer type of sporangia terminal intermingled with smaller type (X 600).

Figs. 36-38, *B. caduca* : 36-habit (X140), 37-sporangia smaller type (X 600), 38-sporangia-longer type (X 600).

Figs. 39-43, *B. coronata* : 39-habit (X 200), 40- a branch (X 600), 41-sporangia-smaller type terminally borne (X 900), 42-sporangia larger, cylindrical (X 900), 43-branches showing axillary sporangia longer ascidium like (X 600).

Figs. 44-45, *B. heterosporangia* : 44- habit (X 200), 45- showing two types and sizes of sporangia (X 600).

Plate 4

Plate - 5

Fig. A.	*Blastocladia pringsheimii* - plant bearing setae, zoosporangia and resting sporangia.
Fig. B.	*B. globosa* - plant with spherical basal cell and zoosporangia.
Figs C&D.	*B. ramosa* - C, habit of plant showing branching and zoosporangia and resting spores. D- a portion of thallus with zoosporangia and resting spores.
Fig. E.	*B. rostrata* - habit (X 72).
Fig. F.	*B. sparowii* - habit with many cylindrical sporangia (X 192).
Figs. G & H.	*Catenaria anguillulae* G. Developing thallus with rhizomycelium H. Mature thallus showing long exit tubes.
Figs. A-D.	Bhargava and Singh (1965).
Figs. E&F.	Das Gupta and John (1953).
Figs. G&H	Misra and Dwivedi (1987).

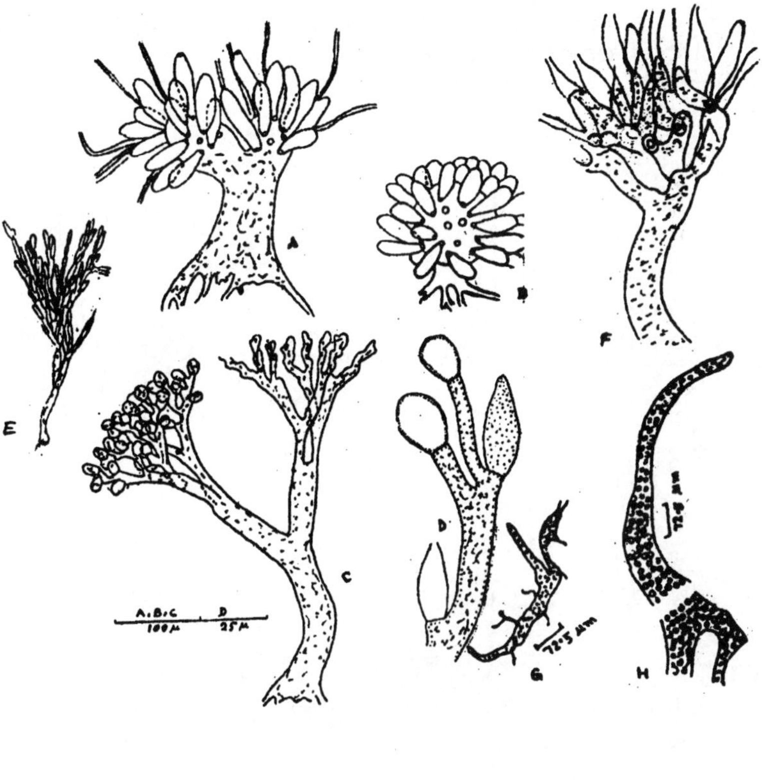

Plate 5

Plate-6

Figs. 1-2: *Blastocladia pringsheimii*: 1. Thallus with thick walled resistant sporangia.

2. Terminal branch of a thallus with ovoid and spherical brown resistant sporangia.

(Photo : Prof. R. Dayal, Varanasi).

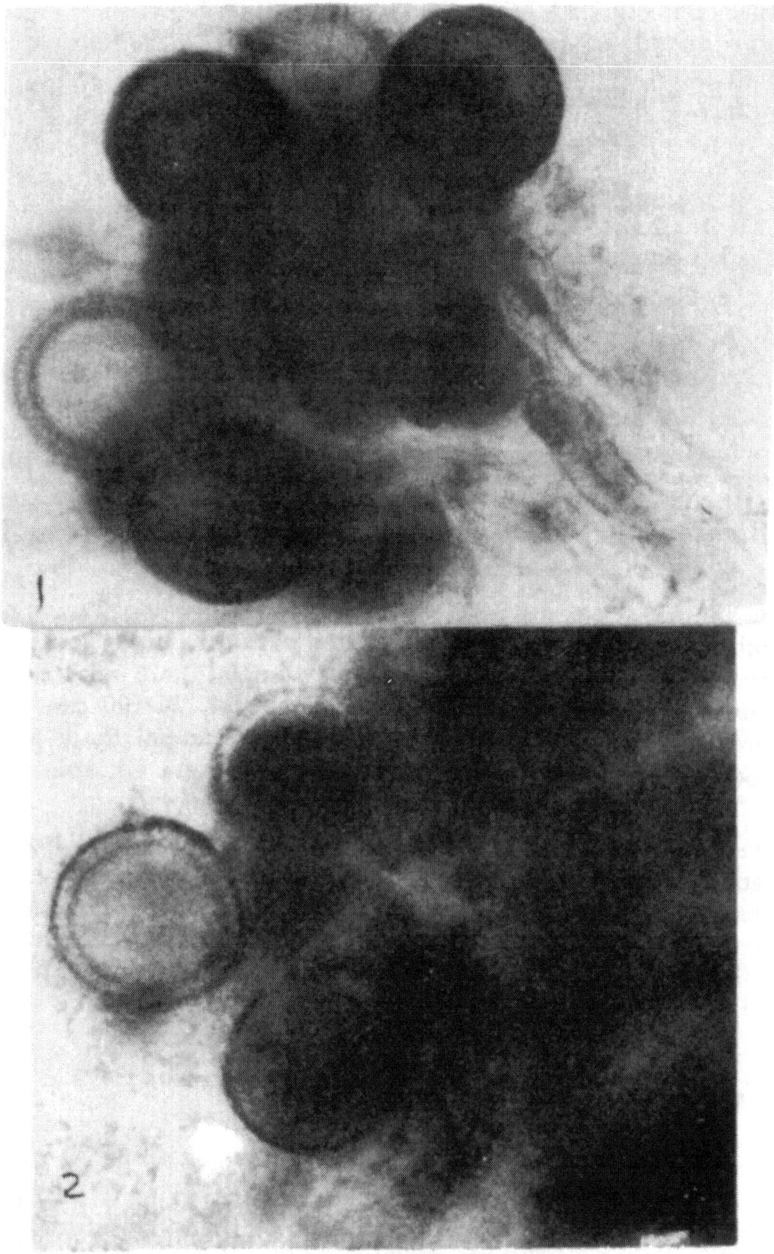

Plate 6

Blastocladiella Matthews

J. Elisha Mitchell Sci. Soc. 53 : 194, 1937

Aquatic Phycomycetes, p. 660, 1960

Key to the species of *Blastocladiella*

1. Gametothalli present .... *B. variabilis,* p. 41
2. Gametothalli absent .... A

A_1 Resting sporangia present
on same thallus .... *B. simplex,* p. 40

A_2 Resting sporangia on
another thallus .... *B. asperosperma,* p. 40

Blastocladiella asperosperma Couch and Whiffen

Amer. J. Bot. **29** : 588; 1942

Thallus extramatrical, rhizoids present, bear sessile zoosporangia and resting spores; zoosporangial thalli spherical, ovoid, pyriforms, subcylindrical to irregular, 20-100 µm in diameter or 50-80 × 100-160 µm; resting sporangial thalli are similar to zoosporangial thalli; resting sporangia variable as zoosporangia (Hasija and Khan, 1983).

Isolated on human nails by Hasija and Khan (1983) from fresh water (Temp. 19°C, pH 7.6) of Engineering College lake of Jabalpur in December, 1978 (Holotype–HC 10, No, 33742). Misra *et al.* (1990) reported it from decaying plant material and bottom sediments of Ganga Sagar polluted lake of Jabalpur during July and September.

Blastocladiella simplex Matthews

J. Elisha Mitchell Sci. Soc., **53** : *194, 1937*

(Plate 7; Figs. 4-7)

Thallus eucarpic, stalked, anchored to the substratum by rhizoid; basal cell cylindrical; holdfast delicate, branched; wall thin, smooth, stalk of the thallus 480-970 µm long by 30-67 µm in diameter; sporangia hyaline, spherical, obpyriform or doliform,

60-150 × 20-50 µm, spherical ones 22-203 µm in diameter or 325 × 135 µm; smooth walled; rhizoidal system well developed, profused; zoospores hyaline, ovoid (5-7 × 4 µm), liberate freely from a single discharge pore; resting spores borne in structure resembling sporangium; thick walled, blackish brown with orange tinge, smooth walled or spiny, 15-140 µm clavate, wall thick, dark brown, basal cell of the resting sporangium 100-150 µm thick; germination of the resting spores not observed (Misra and Dwivedi, 1987).

This species was isolated on flies from water near Medical College, Varanasi (Thakurji, 1967); water of Engineering College lake and Sadar pond, Jabalpur on human nail during July and March, 1979 at water temperature from 28-32°C, pH from 8-8.2 (Khan, 1981); on snake skin from polluted water of Ganga Sagar, Jabalpur (Mishra *et al.*, 1990); on boiled snake skin from lake water and mud soil of Varanasi, wide range of thalli (Mishra and Dwivedi, 1987); in Ganga Sagar lake of Jabalpur during November, August and September (Mishra *et al.*, 1990); from Roberson and Sadar lake (Summer & monsoon) and Hanumantal pond (winter, summer and monsoon), Jabalpur (Agarwal *et al.*, 1989).

Blastocladiella variabilis Harder and Sorgel

Deutsch - Dominik. Tropen. Humburg-Ciudad Trujillo DSD.- 1:123, 1939.

Geobios 5(1) : 36, 1978

(Plate 7; Figs. 8-12)

Hold fast delicate, 30-45 µm in diam., branched (rhizoidal process), thick walled, smooth; sporangia clavate or oval, thick walled, 104-140 µm long 90-100 µm in diam., hyaline with 1-3 distinct discharge papillae, rarely becoming 15-20 µm long discharge tube like structure; zoospores ovoid, rotate prior to discharge, 2.5-3 µm in diam., resting sporangia borne like zoosporangia, spherical with a light yellow to brown colored thick, smooth walled; gametothalli like the sporothalli producing oval to spherical gametangia, slightly smaller than the zoosporangia, 100-130 µm in diam., male gametangia orange colored, while female hyaline, producing isogametes (Prabhuji and Srivastava, 1978a).

Isolated on keratinophilous substrates especially sterile human nails on September 19, 1976 from the soils of a road side puddle near Bus station, Gorakhpur (at pH 7.5, moisture 60 per cent and organic content 4 per cent) (Prabhuji and Srivastava, 1978 a and b); on snake skin from fresh water (30°C, pH 8) of Engineering College lake of Jabalpur during June 1979 by Khan, Specimen HC 18, No. 33740 (f) (Khan, 1981, Hasija and Khan, 1983); Supatal lake, Jabalpur (Indira *et al.*, 1988); from Robertson lake, Jabalpur during monsoon (Agarwal *et al.*, 1989); surface water of decaying material of polluted water of Ganga sagar, Jabalpur, during October to December, March and August to September (Mishra *et al.*, 1990).

Catenaria anguillulae Sorokin

Ann. Sci. Nat. Bot., VI, **4** : 67. 1876

(Plate 5; Figs. G & H)

Thallus predominantly polycentric; zoosporangia hyaline, smooth, usually spherical, oval and sometimes spindle shaped, 60 µm in diam. with 1-3 straight, curved or irregular exit tube; mycelium 5-150 µm long and hyphae 10 µm in diam., zoospores spherical with a single large hyaline refractive globules; rhizoids numerous, well developed and branched, resting spores spherical or oval, 75 µm in diam., occasionally spindle shaped and elongated with a heavy brown wall, 4-8 µm thick, and evenly granular contents.

Parasitic on rotifers, Khandala, Pune, December, 1976 (Kapadnis, 1980); on snake skin bait, lake water, Varanasi; isolate No. AP 11 and 12 (Mishra and Dwivedi, 1987); on snake skin from bottom sediments of Ganga Sagar lake during February to July (Mishra *et al.*, 1990).

Catenaria sphaerocarpa Karling

Amer. J. Bot., **25** : 328, 1938

Saprophytic on *Spirogyra* sp., Khandala, Pune, April, 1976 (Kapadnis, 1980).

Catenomyces Hanson

Torreya, **44** : 30, 1944

Aquatic Phycomycetes, p. 658, 1960

Catenomyces persicinus **Hanson**

Torreya, **44** : 30, 1944

(Plate 7; Figs. 13-16)

Thallus variable in shape and size, eucarpic, polycentric, intra and extramatrical, branched, hyphae unbranched, sterile isthmuses between two sporangia, tubular, triangular, rectangular, cylindrical, elliptical, unbranched and extensive; rhizomycelium extensive, hyphae like, bearing tapering rhizoids or lateral branches forming rhizoidal system; intramatrical hyphae protoplasm with golden refractive globule, extramatrical hyphae, protoplasm hyaline; sporangia intercalary, smooth, peach coloured, pyriform, ovoid a variable in shape, size 22.5 - 150 μm with more than one exit tubes with endo-operculum; tube 78 - 102 μm in length by 7.2 - 7.5 μm in diam., rarely branched, plugged with mucilaginous materials; zoospores spherical, 3.8 - 4 μm in diam. with 2 - 3 refractile globule; resting spores not seen (Gupta and Mehrotra, 1992).

It was collected on snake skin bait from Kurukshetra ponds (Gupta and Mehrotra, 1992). Previously this species was reported by Karling (1966) from Tamil Nadu and Dayal and Kiran (1978) from Varanasi.

Plate - 7

Fig. 1. *Gonapodya prolifera* - Habit and sporangia (X 260) (Das Gupta and John, 1953).

Fig. 2. *G. polymorpha* - Habit with sporangia (X 260) (Das Gupta and John, 1953).

Fig. 3. *G. polymorpha* - Branched hyphae bearing proliferating sporangia, some of them are discharging their spores (X 200). (Bhargava and Singh 1965).

Figs. 4-7, *Blastocladiella simplex* : 4- young sproangium, 5- empty sporangium with single exit pore. 6- zoospore, 7-resting spore showing germination (Mishra and Dwivedi, 1987).

Figs. 8-12. *B. variabilis* - 8- encysted zoospores,
9. habit showing young sporangia.
10. sporangium,
11 and 12 - resting spores.

Figs. 13-16 *Catenomyces persicinus*

13. Immature sporangium with delicate rhizomycelium,
14. Partly discharged mature sporangium with two tubes.
15. A portion of the thallus showing sporangium with several exit tubes and having mucilaginous plug in one of them.
16. Chain of empty intercalary sporangia with tubular isthmuses (Gupta and Mehrotra, 1992).

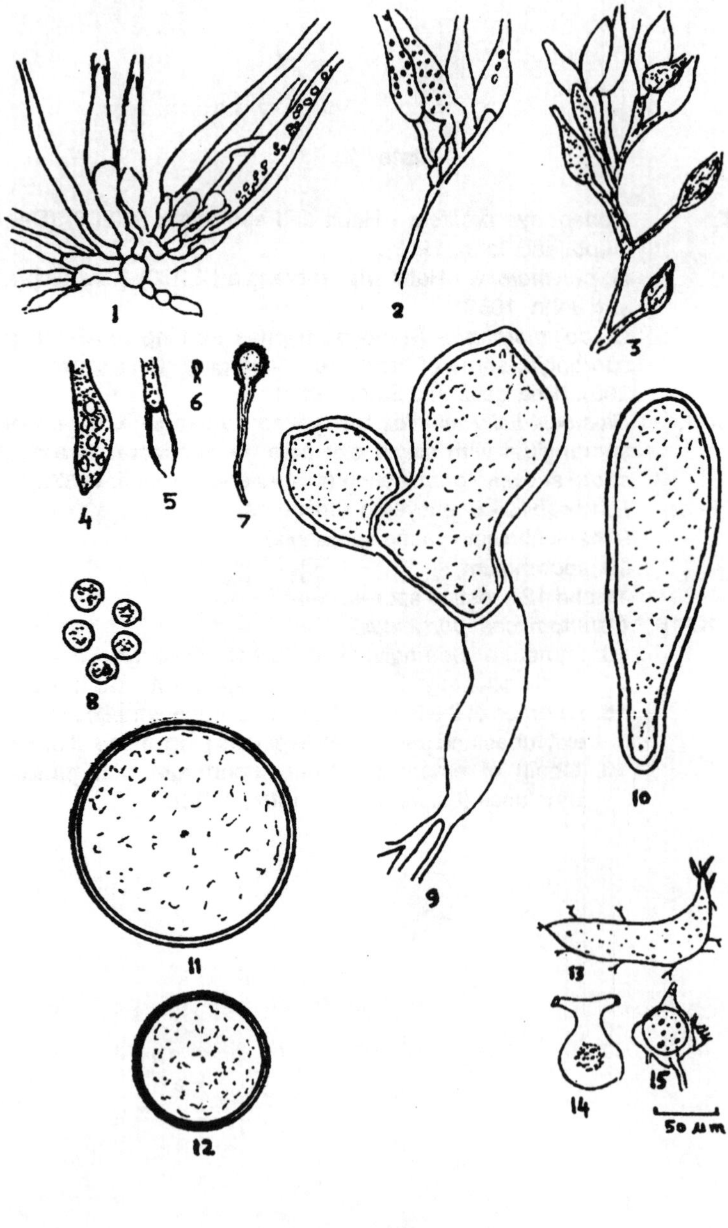

Plate 7

MONOBLEPHARIDALES

1. **Gonapodyaceae :**
   Gonapodya
2. **Monoblepharidaceae :**
   Monoblepharis

Gonapodya Fischer

Rabenhorst Kryptogamen Fl., **1**(4) : 382, 1892
Aquatic Phycomycetes, p. 715, 1960

Key to the Species of *Gonapodya*

1. Mycelium sparingly branched, sporangia with blunt tip .... *G. polymorpha,* p. 49
2. Mycelium profusely branched, sporangia with very elongate distal portions............... *G. prolifera,* p. 50

Gonapodya polymorpha Thaxter

Bot. Gaz. **20**: 481, 1895
Proc. Indian Acad. Sci. **38**(B): 165-170, 1953
(Plate 7, Figs. 2-3)

The thallus is filamentous and sparingly branched; hyphae are usually constricted at intervals, and in places pseudoseptae are formed; the segmented portion is sometimes borne on slender hyphae in which segmentation is indistinct or absent; in a few cases hyphae without any constriction; even at the end when sporangia are borne the constrictions seem to be absent; the protoplasmic contents show foamy or reticulate nature; at the end the constrictions become rounded and ovate sporangia are borne on these; sporangia vary from 50-100 µm in length by 18-30 µm in diam., terminal and solitary or forming fascicles; they do not have elongate distal portion like *G. prolifera*, but taper abruptly to a blunt tip; they are one to many times proliferous; zoospores many in a single sporangium, similar to those of *G. prolifera*; sexual stage not found (Das Gupta and John, 1953).

Isolated on submerged fruits from Lucknow (Das Gupta and John, 1953). Bhargava and Singh (1965) isolated this species from waters of Ramgarh lake, garden pond and Rapti river on a wide range of fruit baits with pronounced preference towards paniyala fruits (*Bischofia javanica*).

Misra and Dwivedi (1987) isolated it on pericarp of decaying Mango fruits submerged in Gujar lake, Jaunpur. His isolate was characterized as follows :

Thallus myceliod, hyphae branched into cylindrical and irregular segments at the crown of branches, segments distinct only to the proximal end of the thallus; sporangia terminal, in fasicles, 30-80 × 25-40 µm proliferating, internal proliferation leads to the projection of growing hyphae beyond the empty sporangium.

Gonapodya prolifera (Cornu) Fisher

Robenhorst. Kryptogamen-F1. **1** (4): 382, 1892

Proc. Indian Acad. Sci. **38** (B): 165-170

(Plate 7, Fig.1)

The mycelium is profusely branched, and the branches diverge from a common base; the hyphae are divided by pseudosepta or constriction into clavate segments; long and pod shaped, an inflated bottom and a very elongate distal portion, and it is the distinguishing feature of this species from *G. polymorpha* ; sessile or separated from the terminal cells by a constriction or pseudosepta; size 70-120 µm long by 15-20 µm in diam., the plants are white to dark brown in colour; zoospores with numerous small granules at the apex; sexual reproduction not found (Das Gupta and John, 1953).

Isolated on submerged fruit from Lucknow (Das Gupta and John, 1953); on apple, grape, tomato and panyala fruit baits in waters of Ramgarh lake and garden pond of Gorakhpur University (Bhargava and Singh, 1965), and from water, Jabalpur (Agarwal *et al.*, 1989). Manoharachary isolated it on apple and mango fruit baits from temple pond of Anatari Hill, Vikarabad and Typha pond, Osmania University Campus, Hyderabad in monsoon and winter seasons during 1971-72, when temperature was low (21.4-26.1°C) and high concentration of nutrients prevailed. He also reported that higher temperature and low concentration of nutrients were the limiting factors. Manoharachary *et al*. (1983) found that this species was associated with some raw mango fruit baiting in pond A, Hyderabad, during winter season (October-January, 1981).

Kiran and Dayal (1992) reported secondary sporangia only slightly exserted, long pod shaped, male and female gametangia one to at least few times proliferous, successive gametangia progressively smaller, female gametangia elongate, ovoid, forming a single discharge papilla, 60-180 × 25-35 µm, female gametes variable in number, up to 20; male gametangia smaller and slender than female, gametes variable in number.

Misra and Dwivedi (1987) isolated it on pericarps of decaying mango fruits placed submerged in Gujar lake, Jaunpur. His isolate showed following characters :

Mycelium dichotomously branched and irregularly divided by hyaline constricted pseudosepta giving an appearance of short elliptical, long, clavate or oval segments (12-20 × 6-8 µm); sporangia terminal, 80-120 × 8-10 µm, proliferating, male and female gametangia rare.

Monoblepharis Cornu

Bull. Soc. Bot. France, *18* : 59, 1871

Aquatic Phycomycetes, p. 726, 1960

Monoblepharis regignens Legerheim

Bih, Kgl. Svensk. Vetensk. AK. Handl., **25.** Afd.3, No. **8** : 39. 1900

Mycelium branched, hyphae thin, 3-6 µm at base,tapering; sporangia cylindrical, 15-40 × 4-8 µm; secondary sporangia formed by internal proliferation of primary sporangia or laterally by cymose manner: sexual reproduction not observed.

This species was isolated from soil and water samples of Lucknow (Chowdhry and Aggarwal, 1980b).

II. OOMYCETES

SAPROLEGNIALES

Saprolegniaceae

Key to the genera of Saprolegniaceae

- Spores monoplanetic, release from the tip,.. swimming 1
- Spores diplanetic, release swimmingly, encyst within or at the mouth 2

la. Sporangia pyriform, bear many zoospores.. *Pythiopsis*, p. 165

lb. Sporangia spherical as oogonia, bear a few zoospores .... *Hamidia*, p. 145

2a Sporangia rare, swimming stage completely suppressed .... *Aplanes*, p.110

2b Sporangia common, swimming stage present...3

3a Primary zoospores encyst within the zoosporangium 4

3b Primary zoospores encyst outside the zoosporangium 5

4a Zoospores arrange in one row, encysted zoospores set free separately by breaking of sporangial wall. .... *Geolegnia*, p. 142

4b Zoospores mostly in many rows, secondary zoospores escape separately through pores formed on sporangial wall. .... *Dictyuchus*, p. 129

4c Zoospores in many rows, encysted primary zoospores set free in a mass by breaking up of the zoosporangial wall at one or more places .... *Thraustotheca*, p. 211

4d. Zoospores in one or many rows, encysted primary zoospores set free by dissolution of zoosporangial wall. .... *Brevilegnia*, p. 114

5a. Primary zoospores strictly in one row 6
5b. Primary zoospores in many rows 7
6a. Sporangia simple or branches, branches not modified into spikes 8
6b. Sporangia branched, branches modified into spiklets, which catch rotifers....................*Sommerstorffia*, p. 205
7a. Primary zoospores escape from the tip of the zoosporangium and encyst at the mouth in spherical or irregular colony .... *Achlya*, p. 57
7b. Zoospores variable in behaviour, some encyst at the mouth as in *Achlya*, and other behave as *Saprolegnia*. Sometimes spores escape from different pores of the zoosporangium. Secondary zoosporangia cymosely as in *Achlya* or internally like *Saprolegnia*. .... *Protoachlya*, p. 158
7c. Primary zoospores escape from the zoosporangium and swim violently 9
8a. Zoospores release one by one and swim violently .... *Leptolegnia*, p. 155
8b. Zoospores swim and encyst at the mouth of the sporangium and form a spherical or irregular colony .... *Aphanomyces*, p.96
9a. Primary zoosporangia long, curved, wavy, branched, papillate; zoospores release through papillae... *Scoliolegnia*, p. 205
9b. Primary zoosporangia small, long, cylindrical, straight, not papillate; zoospores release through apical pores 10
10a. Secondary zoosporangia mostly internal (proliferation) ....*Saprolegnia*, p.169
10b. Secondary zoosporangia mostly lateral.....*Isoachlya*, p.145

Achlya Nees

Nova Acta Acad. Leop. Carol., **11** : 514, 1823

Hyphae slender, stout, straight, tapering, branched, variable in length, 4-188.8 µm in diameter at base; zoosporangia filiform, fusiform, clavate, naviculate, broadest in the middle or towards the base, gradually pointed, 23-1250 µm long and 15-88.4 µm thick; zoospores slightly motile within the zoosporangium, come to rest at once, or after a short period of slow moving, in a hollow sphere or irregular mass at the tip of zoosporangium, encyst there and after a few hours encysted zoospores release; secondary zoospores as in *Saprolegnia* ; encysted primary zoospores 5.5-16 µm in diameter; gemmae present in most of the species, variable in shape, size, position and function (function as zoosporangia or germinate by one or many germ tubes); oogonia mostly lateral on short to long, single, straight or curved stalks, terminal, intercalary, sessile, in some species proliferate, shape variable, mostly spherical or pyriform, 30-180 µm in diameter; oogonial walls thin or thick, double layered, hyaline, yellow, brown, smooth or variously ornamented, spines 3.4-10.2 µm long, pitted (conspicuous or inconspicuous) or unpitted; antheridial branches diclinous, monoclinous, androgynous or exigynous, rarely hypogynous or absent; antheridial cells tubular and clavate, laterally or apically appressed to the oogonial wall; oospores 1-26, centric, subcentric, or eccentric. rarely abortive, spherical, hyaline, yellowish, 14-70 µm in diameter.

In India, 35 species of *Achlya* have been isolated so far. Their identifying key is as follows :

Key to the species

Oospores centric .... A

Oospores eccentric .... B

Oospores subcentric C

A. A_1 Oogonia smooth walled .... D

A_2 Oogonia densely ornamented .... E

B. B_1 Oogonia smooth walled .... G

B_2 Oogonia ornamented .... H

C. C_1 Oogonial wall ornamented .... S
C_2 Oogonial wall smooth .... T
D. D_1 Antheridia absent .... *A. cornuta*, p. 67
D_2 Antheridia present .... F
E. E_1 Antherida mostly androgynous, rarely exigynous .... *A. colorata*, p. 66
E_2 Antheridia mostly monoclinous, rarely androgynous .... *A. papillosa*, p.79
F. F_1 Antherida mostly hypogynous, rarely diclinous and androgynous .... *A. hypogyna*, p.72
F_2 Antherida monoclinous and diclinous, oogonia in racemose manner .... *A. racemosa*, p.83
F_3 Antheridia mostly androgynous, rarely diclinous .... *A. oligacantha*, p.77
F_4 Antheridia androgynous and diclinous, compound .... *A. androcomposita*, p. 62
G. G_1 Antheridia not always present I
G_2 Antheridia always present .... J
H. H_1 Crenulate papillae, antheridia sparse or lacking, if present monoclinous or diclinous .... *A. crenulata*, p. 67
H_2 Truncate ornamentation, oogonial stalk bent, antheridia present, androgynous... *A. recurva*, p. 83
I. I_1 Antheridia diclinous, branched, oogonial wall yellowish .... *A. aplanes*, p. 63
I_2 Antheridia diclinous or androgynous, simple, oogonial wall colourless, oogonia lateral, intercalary, single, or in chain .... *A. de Baryana*, p.68
I_3 Occasional ornamentation, antheridia rare, monoclinous and diclinous, rarely androgynous .... *A. caroliniana*, p.65

J. J_1 Oogonial wall infrequently papillate .... *A. aquatica*, p. 64

J_2 Oogonial wall strictly smooth walled .. K

K. K_1 Antheridial branches wrapping about hyphae or oogonia or both L

K_2 Antheridial branches not wrapping about hyphae and oogonia .... M

L. L_1 A few zoospores motile, not encyst at the tip of sporangium, some oogonia proliferate, oospores 1-12 ....*A. aplanes* var. *indica* p. 64

L_2 Antheridial branches diclinous, twisted, oospores 2-24 .... *A. prolifera*, p.80

L_3 Antheridial branches mostly diclinous, rarely monoclinous, oospores 2-18 .... *A. proliferoides*, p.82

M. M_1 Oogonial wall with few to many conspicuous pits .... N

M_2 Oogonial wall with no pits, except under point of attachment of antheridial cells .... O

N. N_1 Oospores mainly abortive .... P

N_2 Oospheres maturing .... Q

O. O_1 Oogonial stalk curved .... *A. orion*, p. 78

O_2 Oogonial stalk straight .... R

P. P_1 Oospores abort after maturing, antheridial branches diclinous, rarely monoclinous, oogonia like hyphal swellings .... *A. diffusa*, p. 69

P_2 Oospheres rarely maturing, oogonial wall yellowish, antheridia androgynous ..... *A.conspicua*, p.66

Q. Q_1 Antheridia strictly diclinous .... *A. oryzae*, p.79

Q_2 Antheridia mostly diclinous, rarely androgynous .... *A. flagellata*, p.71

Q_3 Antheridia diclinous and androgynous equally present .... *A. imperfecta*, p.73

Q_4 Antheridia androgynous and monoclinous..... *A. americana*, p. 61

R. R_1 Antheridia strictly diclinous, oospore 1- 12 .... *A. klebsiana*, p.74

R_2 Antheridia strictly diclinous, oospore one .... *A.rodrigueziana*, p.84

R_3 Antheridia mostly diclinous, rarely monoclinous .... *A. dubia*, p.69

R_4 Antheridia mostly diclinous, rarely androgynous .... *A. dubia* var. *pigmenta*, p. 70

S. S_1 Antheridia absent or present on 1 per cent oogonia, monoclinous .... *A. stellata*, p.84

S_2 Antheridia present, monoclinous......... A .... *A. stellata var. multispora*, p.85

T. T_1 Oogonia not frequent, antheridia not on all oogonia .... *A.kashyapia*, p.74

T_2 Oogonia frequent, antheridia on all oogonia .... U

U. U_1 Antheridia mostly diclinous, rarely androgynous V

U_2 Antheridia diclinous as well as androgynous .... *A. polyandra*, p.80

U_3 Antheridia mostly monoclinous, very rarely androgynous and diclinous .... *A. apiculata*, p.63

V. V_1 Antheridial cell bulbous, branched, oospore 22-26 µm .... A. *oblongata*, p.76

V_2 Antheridial cell not bulbous, simple, oospore 43-53 µm .... *A. megasperma*, p.76

Achlya americana Humphrey

Trans. Amer. Phil. Soc. **17** : 116, 1892 (1893)

The Saprolegniaceae, p. 111, 1923.

(Plate 9; Figs. 1-3)

Principal hyphae stout, branched, 30-150 µm thick at base, tapering; zoosporangia abundant, long, slender, straight, fusiform, 100-500 µm long by 10-70 µm diameter, predominantly 300 × 50 µm, renewed sympodially; zoospores discharge achlyoid; encysted primary zoospores 7.5 - 11 µm in diameter; gemmae few, elongated, terminal or intercalary, 2-6 in a row; oogonia abundant, borne racemosely from base to tip on main hyphae or from oogonial branches, spherical, rarely pyriform, 40-100 µm in diameter, predominantly 46 - 80 µm; oogonial stalks mostly smaller than the diameter of the oogonia; oogonial wall thin with many conspicuous pits; antheridial branches androgynous and monoclinous, 1 - 7 antheridia per oogonium; antheridial cells elongated, rarely tubular, appressed laterally to the oogonial wall; oospores spherical, eccentric, 1 - 25 in number, generally 5 - 8; size variable even in the same oogonium, ranges from 15 - 30 µm in diameter, mostly 25 µm (Khulbe, 1985).

This species was isolated by P. L. Kochhar (Chaudhuri and Kochhar, 1935) for the first time on boiled house fly, eggs, corn grain from Shalimar and Lahore. Their isolates were characterized by smaller size of zoospores (8.3 - 9.1 µm); oogonia (40.1-68.7 µm, mostly 46 µm) and smaller number of eggs in the normal oogonia (2-6, rarely 13-15) and oospores subcentric to eccentric. Later, this species was reported by Saksena and Rajgopalan (1958), Dayal and Tandon (1962), Srivastava (1967a), Thakurji (1970), Hasija and Khan (1987), Khulbe (1980b), Chowdhry and Aggarwal (1980 a & b), Kapadnis (1980), Mer *et al.* (1980), Misra (1982), Mer and Khulbe (1984), Khulbe (1985), Agarwal *et al.* (1989), Gupta and Mehrotra (1989 a & b) and Mishra *et al.* (1990).

In warmer region of India, it was isolated from water samples of 8 different locations at Jabalpur during 22-28°C temperature, 7.1-7.9 pH, 210 -310 mg/ℓ B.O.D. and 6-12 mg/ℓ Co_2 (Hasija and Batra, 1978); soil and water samples of Kasargod (Chowdhry and Aggarwal, 1980 b); found as common saprophyte in Pune in September, 1976 (Kapadnis, 1980); in surface water, decaying plant materials and bottom sediments in polluted water of Ganga Sagar, Jabalpur during January and March (Mishra *et al* ., 1990).

Temperate region of India also showed occurrence of this species in soils and waters. Mer *et al*. (1980) were the first to report this species from Sat Tal (Naini Tal) lake waters during spring and rainy seasons at 22-24°C temperature. Later, this species was collected from Ram Tal lake (Sat Tal) and found to be characterized by complete absence of androgynous antheridia (Mer and Khulbe, 1984).

This species was again found to be widely distributed in Kumaun Himalaya, viz., lake waters of Khurpa Tal and Naini Tal; forest soils of Almora, Ganai, Ranikhet; cultivated soils of Bhatraunjkhan and Ganai; oak forest soils of Naina Peak (Naini Tal) in all the seasons except winter, and in sal forest soils of Haldwani throughout the year at temperature 5-30°C, moisture 2-90 per cent, pH 5.3-6.9 and organic carbon 0.8-4.9 per cent, (Khulbe, 1985), and in paddy fields of Khatima, Tarai, Nainital (Verma, 1984).

Achlya androcomposita Hamid

Proc. Indian Acad. Sci. XV (4) B: 206-215, 1942

(Plate 11; Fig. 1)

Growth not delicate, hyphae 13.8-41.6 μm, branched; zoosporangia 400-750 μm long, cymosely branched; zoospores 8.3 μm in diameter; oogonia globular, borne racemosely, 61-75 μm in diameter, wall smooth; stalk almost equal in length to the diameter of the oogonia; the number of eggs varies from 4-11, generally 6-8; diameter of eggs 20.8-30.5 μm, mostly 25-27 μm;; centric; antheridia androgynous as well as diclinous; diclinous conditions more abundant; antheridia more than one per oogonium; sometimes arise from the base of the oogonia (Hamid, 1942).

This species was isolated on insect exuvae from Amritsar, Lahore and Hoshiarpur, showing some distinguishing characters, such as unpitted wall of oogonia, compound antheridia, diclinous as well as androgynous antheridia, 20-30.5 μm eggs in diameter, 4-11 eggs per oogonium and long zoosporangia from 400 to 750 μm (Hamid, 1942).

Achlya apiculata de Bary

Bot. Z. **46**: 635, 1888

(Plate 11; Figs. 2-3)

Mycelial growth diffuse and abundant; principal hyphae slender, sparingly branched, 36-88 µm in diameter at base; the main filaments breaking up into segments, each segment becoming a gemma and later forming spores like zoosporangia; zoosporangia sparse, fusiform or clavate; straight, 155-870 µm in diameter, predominantely 290-303 × 25-27 µm; renewed sympodially; zoospores discharge achlyoid; spore cluster persistant at exit pore; encysted spores 9-13 µm in diameter; oogonia variable in abundance; lateral, occasionally spherical, rarely oval, very rarely irregular; 48-122 µm long by 44-103 µm in diameter, predominantly 55-85 × 55-63 µm; oogonial wall thin, smooth, unpitted or pitted only under point of attachment of antheridial cells, inner surface occasionally irregular; antheridial branches monoclinous, occasionally androgynous, rarely diclinous, sparingly branched; antheridial cells short, tubular, slightly irregular; oospores subcentric, spherical or ellipsoidal, filling the oogonium, 1-7 in number, usually 2-4, 22-33 µm in diameter, predominantly 30-33 µm; germination not observed (Dayal and Thakurji, 1968a).

Isolated from pond water, near Tikari village, Varanasi on 20 October 1966, culture RD 27 (Dayal and Thakurji, 1968a). Gupta and Mehrotra (1988a, 1989 a& b) collected it on hemp and *Sesamum* seed baits from Sanhit Sarovar, Kurukshetra.

Achlya aplanes Maurizio

Flora, *79* : 135, 1894

Proc. Indian Acad. Sci. **2**(2) : 137-154

(Plate 11 : Fig. 4)

Growth dense, stout, 112 cm on egg-yellow; hyphae delicate, fine, scarcely branched, 12.3-27.6 µm thick, wavy; zoosporangia thicker, elongated or narrow, clavate, 31.4-40 × 222-504 µm, occasionally as fine as 19.1 µm; chlamydospores abundant, cut off from below the zoosporangia by transverse septa, therefore in chains, cylindrical; oogonia abundant, borne on short lateral stalks, which vary from half to more than twice as long as the diameter of the oogonium; 45.3-67.5 µm in diam., spherical or slightly ovate; wall smooth, thick and yellowish, eggs 1-6, spherical, 24.5-27.6 µm, eccentric; antheridia not always present, mostly diclinous, branched (Chaudhuri and.Kochhar, 1935).

It was collected on house flies from Amritsar and Gujranwala by Chaudhuri and Kochhar (1935) for the first time. Srivastava (1967a) reported it from Ramgarh lake, Gorakhpur.

Achlya aplanes Maurizio var. *indica* Saksena and Dayal

Proc. Nat. Acad. Sci., *28* (B) I : 49-57, 1958

(Plate 11; Figs. 5-7)

Growth vigorous on hemp seeds; main hyphae stout, usually ending in primary sporangia; spores on leaving the sporangium rest at once in a hollow sphere; sporangia frequently retain the spores; if the spores emerge, they sprout at the mouth and in majority of the cultures they do not have any swimming stage (a few swimming zoospores were always observed); oogonia racemose on short stalks, terminal, spherical, 54-126 µm, wall smooth, fairly thick, without any pits; a young oogonium often proliferates to another and empties itself into it; eggs 1 to 12, mostly 3-8, 27 to 45 µm in diameter, eccentric, germination not observed; antheridia diclinous, always wrapping about the oogonia; they generally proceed from more slender branches that run among the oogonial branches (Dayal, 1958).

New variety was formed on the basis of presence of a few motile zoospores, which were completely non motile in the original species. This variety was isolated on hemp seeds from waters of various ponds of Allahabad (Dayal, 1958).

Achlya aquatica Dayal and Thakurji

Myco. et Myco. Applicata, **38** (1-2): 169-174, 1969

(Plate 11; Figs. 8 & 9)

Mycelium moderate and diffuse, two week old colony 1 cm in diameter; hyphae sparsely branched up to 65 µm in diam. at base; gemmae sparse, cylindrical or clavate and single; zoosporangia abundant, fusiform, clavate, 86-396 µm long by 18-61 µm in diameter, predominantly 90-223 x 25-29 µm, renewed sympodially; zoospores discharge achlyoid, spores cluster not persistent at exit pore, encysted spores 7-16 µm in diameter; oogonia abundant, lateral, infrequently papillate, 29-79 µm in diameter, predominantly 40-54 µm, immature ones freely or infrequently proliferating; oogonial stalk short or long, generally stout, straight or curved; oogonial wall pitted only under the point of attachment of antheridial cells, smooth; antheridial branches monoclinous, occasionally androgynous, infrequently diclinous, profusely branched on some oogonia; oospore eccentric, spherical, 1-12 in number, generally 3-5; 14-25 µm in diameter, predominantly 22 µm; germination of oospore not observed (Dayal and Thakur ji, 1969).

This species was isolated on hemp seeds from pools near Daphi, Varanasi by Thakurji on 15th October, 1965. Culture R.D. 13. Isotype at Central Bureau Voor Schimmelcultures, Baarn, Holland (BS 103:67) (Dayal and Thakurji, 1969).

Achlya bisexualis Coker and Couch

J. Elisha Mitchell Sci. Soc., **42**:207, 1927

Saksena and Rajgopalan (1958) reported it from Saugar for the first time. Later, this species was isolated on hemp seed and *Sesamum* seed baits from Brahma Sarovar, Kurukshetra by Gupta and Mehrotra (1988a & b, 1989a & b).

Achlya caroliniana Coker

Bot. Gaz. **50**, 381, 1910

The Saprolegniaceae, p. 123, 1923

(Plate 11; Figs. 22a & b)

Principal hyphae stout, about 48 µm at the base; zoosporangia cylindrical, 20-30 µm in diameter; zoospores release by several openings; sometimes as *Dictyuchus* ; encysted spores 11-12 µm in diam., mostly about 11.5 µm; oogonia abundant, very small, 24-55 µm thick, mostly 30-37 µm; spherical; wall smooth, rarely with 1-2 papillae, thin, not pitted, terminal or lateral on racemose branches; antheridia absent; oospores 1-4, mostly 1-2, 18.5-23 µm in diameter, eccentric with a large oil globule.

This species was isolated from infected *Channa punctatus* fish from Ramgarh lake, Gorakhpur, during January, 1975, for the first time (Srivastava and Srivastava, 1977).

Achlya conspicua Coker

The Saprolegniaceae, p. 131, 1923

Proc. Indian Acad. Sci. II (2) : 137-154, 1935

(Plate 11; Figs. 10-11)

(Plate 12; Fig. 20)

Growth dense, reaching 1.5 cm. on egg-yellow in distilled water; main hyphae long, stout, up to 145 µm thick, near the base, fine branches as small as 24.6 µm; zoosporangia abundant, broadly cylindrical, 15-54.2 × 117-558 µm, secondary ones abundant, produced in groups or borne in a zig zag manner at long intervals; zoospores 8.9-9.9 µm; gemmae abundant, varying in shape, in chains, terminal and intercalary; oogonia many on short lateral stalks, which are as long as the diameter of the oogonium, spherical or oval, occasionally cylindrical and notched, 58.3-64.5 µm; wall yellowish, not thick, slightly pitted; oospores 3-24, spherical, 17-36 µm; eccentric, androgynous, branched, or monoclinous, rarely diclinous (Chaudhuri and Kochhar, 1935).

This species was isolated on egg yellow in distilled water from Amritsar, Atari, Lahore for the first time by Chaudhuri and Kochhar (1935). They observed oogonia with mature oospores after 20 days. Later, it was collected from Jalwania pond, Gorakhpur (Srivastava, 1967a), Neema lake and Garden pond, Allahabad (Dayal and Tandon, 1962), temporary pond, Varanasi, pH 7.5-8 (Dayal and Thakurji, 1965), Brahma Sarovar and Sanhit Sarovar, Kurukshetra (Gupta and Mehrotra, 1988a), Jalalpur lake, Bareilly-Club pond and Dholapur pond, Bareilly (Khare, 1992), Gomati and Kosi rivers of Kumaun Himalaya (Khulbe, 1985).

Achlya colorata Pringsheim

Sitzungsber, der Akad. der Wissensch. zu Berlin, p. 855, 1882.

The Saprolegniaceae, p. 108, 1923

(Plate 13; Figs. 1-4)

Hyphae 25-35 µm in diameter at base, branched; zoosporangia cylindrical, tapering; zoospores release achlyoid, encysted zoospores 10.5-11 µm in diameter; gemmae formed at the maturity of the culture; oogonia few to many, lateral, stalked, rarely terminal up to 145 µm in diameter; mostly 69-78 µm; oogonial wall yellow, with short, blunt growths, rarely smooth, pitted only under point of attachment of antheridial cell, antheridial branches short, androgynous, rarely exigynous; antheridial cells simple, apically appressed; oospores centric, spherical, 15-35, µm in diameter, predominantly 20-31 µm.

It was isolated on hemp seeds from ditch water near agricultural farm, Varanasi by Thakurji on 28 Oct., 1965 and from permanent pond (pH 8-8.5), temporary pond (pH 7.5-8), cemented pond (pH 7-7.5) of Varanasi during September and February to March (Dayal and Thakurji, 1966).

Achlya cornuta Archer

Quart. J. Mic. Sci. **7**: 126, 1867

The Saprolegniaceae, p. 110, 1923

(Plate 11; Fig.21)

Hyphae moderately branched, 4-49.5 µm thick, coenocytic; zoospores liberation achlyoid; zoospores spherical, 6.6 µm in diam.; oogonia few, spherical or elliptical, smooth walled, unpitted, 46.2-72.6 µm in diam., predominantly 42.9 µm.; oospores vary in number, 1-4, 16.5-29.7 µm in diam., centric; antheridia not observed.

Isolated on human nail (bait) by Hasija and Khan (1983) from fresh water (temp. 31°C, pH 6.3) of Deo Tal lake during August 1981 (Specimen HCIO, No. 33744 (b); on hemp seeds from surface decaying plant material and bottom sediments in polluted water of Ganga Sagar, Jabalpur during January (Mishra *et al.*, 1990).

Achlya crenulata Ziegler

Mycologia **40**: 336, 1948

(Plate 11; Fig. 13)

This species is characterized by spherical to hemispherical oogonia with curved stalks, absence of pits, monoclinous antheridia, crenulate oogonia with diclinous antheridia; papillate oogonia with eccentric oospore, varying number of eggs, some abortive oospores.

Isolated on ant and hemp seeds from a pond behind B.H.U. Press, Varanasi by Thakurji on 15th October 1965, culture R.D. 12 (Dayal and Thakurji, 1969).

Achlya de Baryana Humphrey

Proc. Amer. Phil. Soc. (N.S.), 17:117,1893

Proc. Indian Acad. Sci. II(2B) 147, 1935

(Plate 11; Fig. 14)

Growth very dense, 1-7.5 cm on egg yellow; hyphae strong, 19-26 µm thick, more or less branched or ending in zoosporangia, which are somewhat elongated or occasionally clavate, 27.6-30.6 µm thick; secondary ones not so common; zoospores large, 12.1-14.6 µm; chlamydospores very abundant, of various shapes, mostly in chains or branched, each opening by one or more lateral pores; oogonia abundant, borne on short lateral stalks, or very often intercalary, single or in chain, varying much in forms, normally spherical to sub-spherical, very often of abnormal shape, cylindrical or oval, occasionally double, rarely with one or more papillae; 47-61.4 µm in diam., wall thin and smooth; eggs 2-many, spherical, eccentric, 21.6-30.7 µm; antheridia not always present, diclinous or androgynous, mostly latter.

This species was collected from Amritsar by Chaudhuri and Kochhar (1935) for the first time. Later, Manoharachary (1979c) isolated it from pond mud with 9.3 per cent frequency. His isolate was characterized by wider hyphae up to 80 µm thick at base; zoosporangia cylindrical up to 250 µm long, 20-40 µm broad, oogonia short stalked, intercalary, lateral, terminal; antheridial branches monoclinous; oospores eccentric, 1-2 in number in each oogonium, globose and 20-30 µm in diameter.

Manoharachary and Ramarao (1981) isolated it on hempseeds, maize grain and gram leaves from pond-A Hyderabad, during monsoon (June to September) season. Prabhuji (1984) reported it from soils of Gorakhpur. Bhairabnath and Manoharachary (1985) isolated it from two tanks of Nizamabad (A.P.).

Achlya diffusa Harvey ex Johnson

J. Elisha Mitchell, Sci. Soc. *58* : 29, 1942

The genus *Achlya* , p. 64, 1956

(Plate 11; Fig. 15)

Mycelium moderately extensive, diffuse; two week old colony 2-3 cm in diam., principal hyphae slender, sparingly branched, 39-56 µm in diam. at base; gemmae abundant, filiform, irregular,terminal or intercalary; zoosporangia abundant filiform or clavate, 140-870 µm long by 15-62 µm in diameter, predominantly 480-650 × 15-26 µm, renewed sympodially; zoospores discharge achlyoid; spore cluster persistant at exit pore, encysted spores 10-14 µm in diameter; oogonia abundant, mostly lateral, occasionally intercalary, 33-144 µm in diameter, predominantly 33-78 µm, immature freely proliferating; oogonial wall pitting variable in frequency and size; small oogonia unpitted or pitted under point of attachment of antheridial cells, large oogonia pitted; oogonia like hyphal swelling intermingled with true oogonia; antheridial branches usually present, diclinous, rarely monoclinous; oospheres maturing, then quickly aborting; oospores eccentric, not usually filling the oogonium, 1-25 in number, usually 3-5, 15-36 µm in diameter, predominantly 22-26 µm, germination not observed (Dayal and Thakurji, 1968b).

It was isolated on hemp seeds from a pond near Bus station, Gorakhpur (Srivastava, 1967), on maize seeds from a pool near Sarnath, Varanasi (Dayal and Thakurji 1968b), clayey soils around Jalwania pond, Gorakhpur (Prabhuji and Srivastava, 1977), alkaline water of Lucknow (Misra, 1982), soils of Gorakhpur (Prabhuji, 1984), rotted seedlings of wheat at Khatima, Tarai, Nainital (Verma, 1987), Jalalpur lake, Bareilly (Khare, 1992).

Achlya dubia Coker

The Saprolegniaceae, p. 135, 1923

Proc. Indian Acad. Sci. **2**(2): 146, 1935

(Plate 11; Figs. 16, 17a)

Growth rather dense, 1.5 cm on egg yellow; main threads stout, more or less branched, 21.3-40.7 µm thick, tapering gradually; zoosporangia abundant, terminal, long, cylindrical or somewhat clavate, 30.7-42.9 μm, often thinner, those on finer branches as short as 23 × 270 μm; zoospores 10-11·3 μm; oogonia many, on short lateral branches, which vary in length from less than 1 to $^1/_2$ times the diameter of the oogonia, 30-70.5 μm; wall thin, not pitted; antheridia diclinous, sometimes monoclinous; attached by feet; oospores eccentric, spherical, 3-12 in number, usually 3-6; 15-30 μm in diameter, mostly 18-25 μm.

This species was isolated on house fly from water samples, Lahore, Amritsar and Gujaranwala by Chaudhuri and Kochhar (1935) for the first time.Later, it was collected from different parts of India under tropical climate, e.g., on hemp seeds from Jatashankar pond of Gorakhpur (Srivastava, 1967a); on ants from ditch near Medical College, Varanasi, (Dayal and Thakurji, 1969); soil and water samples around I.A.R.I., New Delhi at 10-25°C temperature, (Chowdhry and Aggarwal, 1980 a); from Lucknow, Tejpur (Chowdhry and Aggarwal, 1980b); on hemp seeds from decaying plant materials in bottom sediments in polluted water of Ganga Sagar of Jabalpur in December at 16.0°C temperature, 7.5 pH, 126.0 per cent total alkalinity (Misra, 1982 and Mishra *et al.*, 1990); on hemp and mustard seed from Brahma Sarovar, Kurukshetra (Gupta and Mehrotra, 1988a, 1989a & b).

One isolate of this species collected by Dayal and Thakurji (1969) showed different characters, viz., pitted oogonial wall, diclinous antheridia, 1-8 oospores per oogonium, irregular thraustothecoid discharge of primary zoospores and, decreasing trend of thraustothecoid liberation during subculturing and dominance of achlyoid discharge.

Achlya dubia var. *pigmenta* Chaudhuri and Kochhar

Proc. Indian Acad. Sci. 2(2B): 147, 1935

(Plate 13; Figs. 5 & 6)

Growth slightly dense, hyphae stout, reaching 1 cm. on egg yellow; hyphae 27.6-36.8 μm, branched, straight; zoosporangia thicker, 36-67 × 370-720 μm long, cylindrical, outlines regular and straight; secondary ones arising by cymose branching from below the primary ones; chlamydospores abundant, produced in groups of various shapes; zoospores coming out occasionally in groups and collecting at the tip; mostly retained within the zoosporangium and germinating there; oogonia plentiful, on short lateral stalks, which are up to as long as the diameter of the oogonium; diameter 39.2-62 µm, wall smooth, thin, yellowish or dark; eggs 2-9, usually 3-4; 21.4-24.5 µm in diam. antheridia diclinous or androgynous, mostly former, branched, attached by feet to the oogonial wall.

This species was collected from Lahore and Gujaranwala by Chaudhuri and Kochhar (1935) for the first time. They found small sized oogonial stalks and smaller eggs than those of A. *dubia*. It was also characterized by chlamydospores, coloured oogonial wall and constantly occurring diclinous antheridia.

Dayal and Thakurji (1969) remarked that the identity of *Achlya dubia* var. *pigmenta* was doubtful and the description of variety *pigmenta* was similar to that of *A. klebsiana*.

Achlya flagellata Coker

The Saprolegniaceae, p. 116, 1923

(Plate 9; Fig. 4)

Principal hyphae stout, branched, tapering, 15-120 µm thick at base; zoosporangia abundant, subcylindrical, fusiform, 200-750 µm long by 15-80 µm in diameter, predominantly 300 × 50 µm, renewed sympodially; zoospores discharge achlyoid; encysted primary zoospores 10-11.5 µm in diameter, mostly 10.5 µm; zoosporangia abundant, elongated to fusiform, 135-605 µm in length and 18-37.5 µm in width, mostly 345-490 × 22.5-30 µm, renewed sympodially; zoospores discharge achlyoid; encysted zoospores spherical, 9.5-10.5 µm in diameter; gemmae abundant, single or in chain of 2-5, clavate, behave as zoosporangia, rarely as oogonia; oogonia abundant, lateral, spherical to pyriform, 33.5-75 µm in diameter, predominantly 40-55 µm; oogonial wall thin, smooth, with many conspicuous pits; antheridial branches diclinous, long and branched, 1-3 antheridia appressed per oogonium; antheridial cells tuberous, elongated with short tube like projections; oospores spherical, eccentric, 1-7 per oogonium, generally 3-5, size 16.75-23 µm in diameter, mostly 18.5 µm.

This species was isolated on house fly from Amritsar, Lahore and Gujaranwala by Chaudhuri and Kochhar (1935) for the first time.

Later, this species was reported from different ecological conditions. In tropical region, it was isolated on hemp seeds from pond waters of Allahabad (Dayal, 1958); Neema lake and some ponds of Allahabad (Dayal and Tandon, 1962); permanent, temporary and cemented ponds (pH 7-8.5) of Varanasi during August, September, November, December, February and April (Dayal and Thakurji, 1966); from clayey soils around Jalwania pond (pH 8.5, moisture 35 per cent and organic carbon 4.2 per cent) of Gorakhpur (Prabhuji and Srivastava 1977 a & b); on hemp seeds from Maidagin pond water, Varanasi by Usha Kiran (Kiran and Dayal, 1983).

In colder habitats, this species was isolated on hemp seeds from Naini Tal lake during rainy and autumn seasons, from Khurpa Tal, Bhim Tal and Naukuchiya Tal lakes in autumn when temperature ranged between 15-23.5°C (Khulbe and Bhargava, 1977), pH 5.6-8.8 (Khulbe, 1980 a & b); from rhizosphere and non-rhizosphere soils of tomato, pea and wheat fields near Naini Tal (Khulbe, 1983); from soils of Gorakhpur (Prabhuji, 1984), Haldwani, Naini Tal, Kausani and Pithoragarh, and agricultural soils of Berinag; and lake water of Naini Tal and Sat Tal (Khulbe 1985). This species was found as a regular member of soil mycoflora in oak forest at Naina Peak, Naini Tal during summer and autumn, and in sal forest, Haldwani during spring, rainy and autumn seasons at 14-30°C temperature, 2-50 per cent moisture, 5.3-6.9 pH and 1.8-4.9 organic carbon (Khulbe, 1985 & 1991).

It was also isolated from cultivated paddy field soils and rotted wheat seedlings from Khatima, Tarai, Naini Tal (Verma, 1984, 1987). Gupta and Mehrotra (1988a, 1989a & b) isolated it on hemp and Sesamum seed baits from Sanhit Sarovar, Kurukshetra. Khare (1992) found it as a most frequent watermold of river, lake and pond of Bareilly.

Achlya hypogyna Coker and Pemberton

Bot. Gaz. **45** : 194, 1908

(Plate 11; Fig. 17b)

Mycelium branched moderate, stout, tapering towards the apex, 36-188.8 µm at the base; gemmae moderate, abundant and variable in shape, often in chain and rod shaped formed by the segmentation of the hyphae; sporangia cylindrical or clavate, renewed sympodially, 187.6-522.1 × 67-87.1 µm; zoospore discharge achlyoid, spore cluster irregular and persistant, encysted spores 6.9-8.3 µm in diam.; oogonia on short lateral branches, globular or very rarely oblong., 47.6-74.6 µm in diam.; oogonial wall unpitted and smooth; antheridia hypogynous, diclinous, occasionally androgynous, usually 2-3 antheridia per oogonium, antheridial cells tubular or clavate, laterally appressed to the oogonium; oospores spherical, 1-10 in number, but usually 2-5, centric, 18.7-27.2 µm in diameter, mostly 23.8 µm.

It was isolated from alkaline ponds, Lucknow (Misra, 1982) (pH 8.5); Devi Khera near Lucknow (Rai and Misra, 1976); on hemp and mustard seed baits from Brahma Sarovar, Kurukshetra (Gupta and Mehrotra, 1988a, 1989a & b).

Achlya imperfecta Coker

The Saprolegniaceae, p. 118, 119, 1923

(Plate 11; Fig. 18)

Principal hyphae stout, branched, tapering, 45-85.5 µm thick at base; sporangia numerous; fusiform or naviculate, occasionally filiform or cylindrical; 180-610 µm long by 20-52.5 µm thick, predominantly 45 × 450 µm; renewed sympodially; zoospores discharge achlyoid; encysted spores 8.2-11.5 µm thick in diameter, predominantly 10 µm; gemmae abundant; terminal or intercalary; shape variable, oval to fusiform or irregular; single or in chain of 2-6; mostly resistant, rarely deciduous, dark black; oogonia abundant; spherical or oval; 30-90 µm in diameter; lateral on stalks which are half to four times longer than the diameter of the oogonium; straight, curved; oogonial wall smooth, thin with few to many inconspicuous pits; antheridial branches many; simple or branched; diclinous as well as androgynous; 1-4 antheridia on every oogonium; antheridial cells small, tubular; simple or branched; laterally appressed, rarely attached by projections to the oogonial wall; oospore eccentric; spherical; 1-10 in number, mostly 4-8; 13.2-24.4 µm in diameter, predominantly 23.1 µm.

This species was isolated on house fly baits from Amritsar and Lahore by Chaudhuri and Kochhar (1935) for the first time. Later, it was found to be distributed in warmer waters of some temporary and cemented ponds (pH 7.5-8) of Varanasi during December and March (Dayal and Thakurji, 1966); and in colder waters of Nainital, Khurpatal, Bhim Tal and Naukuchiya Tal lakes during autumn when temperature and pH ranged from 15-23.°C (Khulbe and Bhargava, 1977) and 7.3-8.1 (Khulbe 1980 a & b), respectively. Gupta and Mehrotra (1988a, 1989a & b) isolated it on hemp and mustard seed baits from Sanhit Sarovar, Kurukshetra. Khare (1992) found it in Bareilly-club pond and Dhopeswar pond, Bareilly.

Achlya kashyapia Chaudhuri and Kochhar

Proc. Indian Acad. Sci. **2** (2): 137-154, 1935

(Plate 11; Fig. 19)

Growth rather open, hyphae stout, near the substratum, giving out slender branches, 12.3-18.4 μm, occasionally as wide as 30 μm and as fine as 9.2 μm; zoosporangia plentiful, elongated, cylindrical or sub-cylindrical, more or less smooth, secondary ones quite common and produced as in other forms of this genus, vary variable in size, 32-45 × 252-873.7 μm, occasionally as long as 1 mm., those on fine branches as small as 15.4 × 175 μm; chlamydospores formed usually by the segmentation of hyphae and therefore in chains and cylindrical to subcylindrical, rarely otherwise, not getting loosened and separated from each other; zoospores 9.8-10.5 μm; oogonia not so frequent, when present scattered in groups of 3 or 4, on short lateral stalks, dark coloured, neck short or absent, spherical occasionally flask shaped; wall moderately thick, 2.8-3.1 μm, unpitted, 58.2-76.3 μm, eggs 3-8 in number, 24.6-27.8 μm, spherical, subcentric or centric; antheridia not on all oogonia, diclinous, branching, a single one sending branches of 2 or more.

It was collected from Lahore and Khasa near Amritsar by Chaudhuri and Kochhar (1935) for the first time. They observed diclinous antheridia, unpitted oogonial wall and larger eggs.

Achlya klebsiana Pieters

Bot. Gaz. **60** : 486, 1915

(Plate 9; Figs. 5-6)

Main hyphae moderately stout, 15-70 μm thick at base, branched, tapering; zoosporangia numerous, long, slender or fusiform, 150-700 μm long by 15-80 μm in diameter, predominantly 275 × 25 μm, renewed sympodially; zoospores discharge achlyoid, encysted primary zoospores 9-12 μm thick, mostly 9 μm; gemmae abundant, terminal or intercalary, spherical, cylindrical or rod shaped; oogonia abundant, lateral on short to long stalks, spherical, rarely pyriform or irregular, 30-90 μm in diameter, predominantly 48 μm; oogonial wall smooth, thin, pitted only under the point of antheridial attachment; antheridial branches many, branched, diclinous, 1-3 antheridia per oogonium; antheridial cells elongated; oospores spherical, eccentric, 1-12 in number, mostly 5-8, 15-30 μm in diameter, generally 21.5 μm (Khulbe 1985).

Isolated on ants from Haldwani forest soil on April 15, 1980, leg. R.D. Khulbe. The type culture has been deposited in the Department of Botany, Kumaun University, Nainital (Ach/3). The present isolate was found to be distributed in the forest soils of Nainital and Ramnager.

This species was isolated on house fly baits from Shalampur, Mughalpur, Lahore (Chaudhuri and Kochhar, 1935); Calcutta (Chaudhuri and Lotus, 1936); Gauhati (Bhattacharya and Baruah, 1953); lake water, Hyderabad (Manoharachary, 1979 b); soil and water samples, I.A.R.I., New Delhi (Chowdhry and Aggarwal, 1980a); Haldwani (Chowdhry and Aggarwal, 1980b); pond waters, Hyderabad (Manoharachary and Rao, 1987); cultivated soils of paddy fields and diseased seedlings (Verma, 1984); soils, Gorakhpur (Prabhuji, 1984); tank waters, Nizamabad (Bhairabnath and Manoharachary, 1985); soils around Nal Damyanti Tal; seedlings of *Oryza sativa* sal forest soils, Haldwani, Ramnagar; oak forest soils, Nainital (Khulbe, 1985); Brahma sarovar and Sanhit sarovar, Kurukshetra (Gupta and Mehrotra, 1988, 1989a & b); and Jalalpur lake and Bareilly-club pond, Bareilly (Khare, 1992).

Achlya klebsiana Pieters var. *Kashyapia* Chaudhuri

Proc. Indian Acad. Sci. II; 2B, 145, 1935

It was isolated from Khasa (near Amritsar), Lahore (Chaudhuri and Kochhar, 1935).

Achlya megasperma Humphrey

Trans. Amer. Phil. Soc. *17* : 118, 1892
The Genus *Achlya* p. 36-37, 1956
(Plate 12; Fig. 3b)

Mycelium well developed, main hyphae stout, branched, 35-120 µm at base; gemmae abundant, variable in shape and size, function as zoosporangia; zoosporangia abundant, filiform, clavate, fusiform, 90-1250 × 60-78 µm, mostly 180-500 × 25-40 µm; renewed sympodially, cymosely or in basipetal succession; zoospores discharge achlyoid, encysted spores 10-12 µm in diameter; oogonia abundant, lateral, rarely terminal, spherical, occasionally oval, rarely apiculate, 40-180 µm, mostly 65-80 µm in diameter, oogonial wall unpitted, irregular on inner surface; antheridial branches diclinous, occasionally androgynous, branched, oospores subcentric, spherical, 1-15 in number, generally 2-8, 35-70 µm in diameter, mostly 35-50 µm. This species is characterized by large oospores.

It was isolated from soil and water samples from Kasargod (Chowdhry and Aggarwal, 1980 b) for the first time.

Achlya oblongata de Bary

Bot. Zeit. **46** : 646, 1888
(Plate 12; Fig. 1)

Mycelium moderate, stout and branched, gemmae present, formed by the segmentation of the hyphal chain; sporangia abundant, fusiform, measuring 259-666.6 × 30-77.7 µm, mostly 188-232 × 29-33 µm; renewed sympodially; zoospores discharge achlyoid, spores clusters persistant; encysted spores 5.5-11.08 µm in diameter; oogonia abundant, large, borne laterally on main branches, rarely intercalary, oblong to subspherical, 49.8-122 µm, mostly 55-108 µm, wall smooth, thin; antheridia diclinous, one to many per oogonium, laterally appressed; antheridial cells branched, bulbous; oospores 1-16 in number, subcentric, 18-39 µm in diameter, predominantly 22-26 µm, germination not observed.

The species was isolated from Lahore in April 1936 by Hamid (1942) for the first time. Later, it was collected from pond water of Lacchi Tara, Teli bagh area (pH 7.6), Lucknow (Rai and Mishra,

1977b, Misra 1982); Deo Tal lake, Jabalpur during August 1978 and November 1978 (Hasija and Khan, 1983).

Dayal and Thakurji (1968) isolated it from pond water near Tikari village, Varanasi on 20 Oct. 1966, slide RD 23, and described their isolate as *A. oblongata* var. *oblongata* due to presence of short oogonial stalks and abundant, much branched antheridial branches. Gupta and Mehrotra (1988a, 1989 a & b) reported it from Sanhit pond of Kurukshetra.

Achlya oligacantha de Bary

Bot. Zeit. **46** : 647, 1888

(Plate 12; Fig. 3a)

Main threads slender and delicate; oogonia on short or long branches of hyphae, which are furnished with sporangia or in part terminal on slender main threads and their racemose branches, spherical in shape, varying from 34-88.4 µm in length and 27.2-81.6 µm in breadth, surface always with relatively large, smooth papillae, measuring about 3.4-10.2 µm in length, separated from each other and which vary greatly in number (1-7, very seldom none), also variable in size and form (short points to large, blunt projections); walls of oogonia relatively thin, colourless, without pits except that the projections are mostly thinner than the walls between them; eggs centric, mostly 3-5 in an oogonium (seldom up to 9 or more), round, relatively small, 17-34 µm in number; antheridia always present, mostly several on each oogonium, relatively small; borne one or two in a row, some are androgynous, while other partly diclinous in origin (Dayal and Thakurji, 1965).

It was isolated from water samples of Varanasi (Dayal and Thakurji, 1965 and Thakurji, 1967), Jabalpur (Aggarwal *et al.*, 1989, Khan, 1981) and Kurukshetra (Gupta and Mehrotra, 1988a). Dayal and Thakurji (1966) reported occurrence of this species in temporary pond of Varanasi at pH 7.5-8. Gupta and Mehrotra (1988a, 1989a &b) isolated it on hemp and mustard seed baits from Brahma Sarovar, Kurukshetra. Hasija and Khan (1983) collected it on wheat and hemp seed baits from fresh water (temp. 24-31°C, pH 6.8-8) of Deotal lake during August 1978 and November 1979. He grew it on Miller's medium (type culture-HCIO No. - 33744, Jabalpur University).

Achlya orion Coker and Couch

The Saprolegniaceae, P. 112, 1923

(Plate 9; Figs. 7-9)

Principal hyphae stout, branched, tapering, 7.5-82.5 µm thick at base; zoosporangia abundant, cylindrical, fusiform, 60-740 µm long by 11.2-59 µm in diameter, predominantly 125-360 × 35-55 µm; zoospores discharge achlyoid, encysted primary zoospores 9-12 µm thick in diameter, predominantly 10 µm; gemmae abundant, cylindrical, 150-250 µm long by 20-45 µm thick, behave as zoosporangia; oogonia abundant, spherical, lateral, 25-75 µm in diameter, mostly 40-60 µm; oogonial stalk half to three times longer than the diameter of oogonia, mostly curved down; oogonial wall smooth, without pits; antheridial branches on all oogonia, monoclinous and androgynous; oospores spherical, eccentric, 1-14 in number, mostly 2-4, 19.8-28 µm in diameter, mostly 26-28 µm (Khulbe, 1985).

This species was isolated from Allahabad (Dayal and Tandon, 1962), Varanasi (Thakurji, 1967; Dayal and Thakurji, 1968a &b), Gorakhpur (Srivastava, 1967), Delhi and Haldwani (Chowdhry and Aggarwal, 1980b) and Pune (Kapadnis, 1980).

In warmer habitats, this species was collected from different ecological conditions, viz., from silty soils of Ramgarh lake (pH 7.0, moisture 25 per cent, organic contents 4.5 per cent) of Varanasi (Prabhuji and Srivastava, 1977b); pond water near Chiraigaon, Varanasi (Dayal and Thakurji, 1968a & b); pond water of Vikarabad, which was rich in nutrition and biological population (Manoharachary, 1977); pond water of Hyderabad during winter when dissolved oxygen, oxidizable organic matters, nitrates and phosphates were high and temperature was low (Manoharachary, 1979b); soil and water samples in and around I.A.R.I., New Delhi during July to October when temperature ranged between 20-30°C (Chowdhry and Aggarwal, 1980b); from pond A, Hyderabad during winter (October to January, 1981) (Manoharachary and Rao, 1981); soils of Gorakhpur (Prabhuji, 1984); tank waters, Nizamabad (Bhairabnath and Manoharachary, 1985), and Jalalpur lake and Dhopeswar pond, Bareilly (Khare, 1992).

Temperate habitats also showed isolates of *Achlya orion* at certain places, viz., soils of Ram Tal, Garur Tal and Nal Damyanti Tal (Sat Tal) during summer to rainy seasons at 22.8-24.8°C (Mer *et al.*, 1980); rhizosphere and non-rhizosphere soils of tomato, paddy, pea and wheat near Naini Tal (Khulbe, 1983b); soil samples of Ram Tal, Sat Tal (Mer and Khulbe, 1984); pine forests soils of Almora, Dwarahat, Nainital, and oak forest soils of Naina peak during autumn, and in sal forest soils of Haldwani throughout the year under temperature 14-30°C, pH 5.3-6.8, moisture 2-50 per cent and organic carbon 0.8-4.9 per cent (Khulbe, 1985).

Achlya oryzae Ito et Nagai

Jr. Fac. Agric. Hokkaido Imper. Univ., **32**, 17:1931

(Plate 10; Figs. 1-7)

Main hyphae moderately stout, branched, gradually tapering towards the end, 25-44 µm in width at the base; zoosporangia abundant, elongated to fusiform, 135-605 µm in length and 18-37.5 µm in width, mostly 345-490 × 22.5-30 µm, renewed sympodially; zoospores discharge achlyoid; encysted zoospores spherical, 9.5-10.5 µm in diameter; gemmae abundant, single or in chain of 2-5, clavate, behave as zoosporangia, rarely as oogonia; oogonia abundant, lateral, spherical to pyriform, 33.5-75 µm in diameter, predominantly 40-55 µm; oogonial wall thin, smooth, with many conspicuous pits; antheridial branches diclinous, long and branched, 1-3 antheridia appressed per oogonium; antheridial cells tuberous, elongated with short tube like projections; oospores spherical, eccentric, 1-7 per oogonium, generally 3-5, size 16.75-23 µm in diameter, mostly 18.5 µm (Khulbe, 1985).

Isolated from infected seedlings of *Fagopyrum esculentum* Moench. from Bhim Tal on July 15, 1982, leg. R.D. Khulbe. The type culture has been deposited in the Department of Botany, Kumaun University, Nainital (Ach/5).

Verma and Khulbe (1983) isolated it as a parasite of rice seedlings from Tarai, Nainital.

Achlya papillosa Humphrey

Trans. Amer. Phil. Soc. *17* : 125, 1892

(Plate 12; Fig. 11)

Hyphae slender, long, branched, mycelium extensive, 25-60 µm thick at base; gemmae abundant, variable; zoosporangia rare, filiform, 200-700 × 15-50 µm, predominantly 300-400 × 15-30 µm, renewed sympodially or basipetally; zoospore discharge achlyoid, rarely aplanoid, encysted spores 10-12.5 µm in diameter; oogonia abundant, lateral, rarely terminal, varying in shape, spherical rarely dolifirm, 40-200 µm diameter; oogonial wall unpitted, papillate or tuberculate; antheridial branches abundant, monoclinous, rarely androgynous, branched, 1-5 on an oogonium; oospores centric, spherical, 1-20 in number, mostly 4-5, 15-40 µm in diameter, mostly 25-30 µm.

It was isolated as saprophyte on hemp seeds at Mahabaleswar in September, 1976 (Kapadnis, 1980).

Achlya polyandra Hildebrand

Johrb. F. Wiss. Bot. **6**: 258, 1867.

(Plate 12; Fig. 4)

Mycelial growth dense, hyphae stout; gemmae present, formed by segmentation of the hyphae; sporangia abundant, renewed by cymose branching, 221-502 × 20-53.6 µm, commonly curved; encysted spores 8-11 µm in diameter; oogonia abundant, borne laterally on the main branches, 50.4-108.8 µm in diameter, oogonial wall smooth, unpitted, oogonial stalks little curved in some cases; antheridia both diclinous and androgynous, one to many per oogonium, short and clavate; oospores 2-26 in number; commonly 4, measuring 16.6-28.9 µm in diameter, centric and subcentric.

Isolated from pond water (pH 8.0) near District Jail, Lucknow by Rai and Mishra (1977b) and Misra (1982). Butler (1907) collected it for the first time on decomposing insect under water, Pusa (Chaudhuri *et al.*, 1947).

Achlya prolifera Nees von Esenbeck

Nova Acta Acad. Loop. Carol. **11**: 514, 1823

Bot. Zeit. **10**: 473, 1852

The Saprolegniaceae, p. 143, 1923

(Plate 12; Fig. 5)

Principal hyphae stout, branched, tapering, 20-60 µm thick at base; zoosporangia abundant, cylindrical, 180-780 µm long by

22-75 µm in diameter, predominantly 350-500 × 40-65 µm, renewed sympodially; zoospores discharge achlyoid, encysted primary zoospores 8-11.5 µm in diameter, mostly 10 µm; gemmae abundant, variable in shape, size and position; oogonia abundant, lateral, terminal, intecalary, pyriform, 50-200 µm, mostly 70-100 µm in diameter, smooth, pitted; antheridial branches diclinous, twisted, branched, wrapping about the oogonia; antheridial cells elongated, rarely intercalary; oospores spherical, eccentric, 2-20 in number, mostly 4-8, 20-35 µm in diameter, mostly 25-30 µm (Khulbe, 1985).

This species was reported from Amritsar by Chaudhuri and Lotus (1936) first of all. Later, it was isolated from water samples of Varanasi (Dayal and Thakurji, 1965), temporary and cemented ponds (pH 7.5-8) of Varanasi during December and March (Dayal and Thakurji (1966); Ramgarh lake of Gorakhpur (Srivastava, 1967); silty soils of the environs of Ramgarh lake (pH 6.5, moisture 25 per cent, organic content 4.5 per cent) of Gorakhpur (Prabhuji and Srivastava, 1977); soil and water samples around IARI., New Delhi during all the seasons at 10-45°C temperature (Chowdhury and Aggarwal, 1980a) and soil samples of Rampur, Haldwani, Lucknow, Bangalore, Mangalore ,Ooty, Coonoor (Chowdhry and Aggarwal, 1980b); soils, Gorakhpur (Prabhuji, 1984) from Brahma Sarovar, Kurukshetra (Gupta and Mehrotra, 1988a, 1989a & b). Khare (1992) isolated it from Jalalpur lake, Bareilly club pond and Badaun road pond of Bareilly on decaying plant materials in polluted water of Ganga Sagar, Jabalpur during November, December, July to September (Mishra *et al.*, 1990).

In Kumaun Himalaya, this species was isolated from the soils of Ram Tal (in spring), Nal Damyanti Tal (spring to autumn), Sat Tal at 19.8-26.8°C temperature (Mer *et al.*, 1980); from cold water fishes (Sati, 1981), pine forest soils of Ranikhet and oak forest of Kilbury, Nainital, Naina peak during summer and autumn, and in sal forest soil of Haldwani during spring, summer, rainy and autumn at 14-30°C temperature, 5.3-6.9 pH, 2-50 per cent moisture, 0.4-4.9 per cent organic carbon (Khulbe, 1985); paddy fields of Khatima and Nainital (Verma, 1984).

Achlya proliferoides Coker

The Saprolegniaceae, p. 115, pl. 36, 1923

(Plate 12; Fig. 6)

Principal hyphae stout, frequently branched; gemmae abundant, filiform or irregular, usually intercalary and catenulate; sporangia abundant, filiform or fusiform, usually curved, 150-750 µm long by 25-50 µm in diameter, renewed sympodially or rarely in basipetalous succession; oogonia abundant, usually lateral, very rarely intercalary, spherical or pyriform, sometimes oval, 35-90 µm in diameter, predominantly 45-70 µm; oogonial wall smooth, pitted; antheridial branches mostly diclinous, sometimes monoclinous, usually coiling about hyphae, which may or may not bear oogonia, irregular, much branched; wrapping about the oogonium; antheridial cells tubular, simple or branched, clasping the oogonium; attached by projection, rarely appressed; oospores eccentric, spherical, 2-18 in number, generally 3-8; 20-35 µm in diameter, mostly 25-30 µm (Srivastava, 1967a).

Chaudhuri and Kochhar (1935) reported this species from Lahore and Gujranwala for the first time. Later, it was isolated from Varanasi (Dayal and Thakurji, 1965), permanent pond (pH 8-8.5) and temporary ponds (pH 7.5-8) of Varanasi during January to April (Dayal and Thakurji, 1966), Jalwania Pond, Gorakhpur (Srivastava, 1967a); from clayey soils in the vicinity of Jalwania pond (pH 8, moisture 35 per cent, organic content 4.2 per cent) of Gorakhpur (Prabhuji and Srivastava, 1977); ponds waters of Hyderabad during monsoon season when albuminoid ammonia and free ammonia were at their peak (Manoharachary, 1979b). Chowdhury and Aggarwal (1980 a & b) isolated it from soil and water samples of I.A.R.I., New Delhi, and found to be distributed in Delhi, Rampur and Tejpur throughout the year during 10-45°C air temperature.

Again, this species was reported from some ponds of Hyderabad during monsoon-July to September (Manoharachary and Ramarao, 1981), Varanasi (Kiran and Dayal, 1983), paddy fields of Nanakmatta, Nainital (Verma, 1984), soils of Gorakhpur (Prabhuji, 1984); Brahma Sarovar, Kurukshetra (Gupta and Mehrotra, 1988a, 1989a & b); rivers ponds and lake waters of Bareilly (Khare, 1992).

Achlya racemosa Hildebrand

Jahrb. f. Wiss. Bot. *6* : 249, 1867

(Plate 12; Fig. 7)

Mycelial growth moderate, hyphae thick, gemmae not formed even in old culture; sporangia cylindrical or clavate, 222.2-444.4 µm in diameter, renewed by cymose branching, encysted spores 6-8.31 µm in diameter; oogonia abundant, borne laterally in racemose manner on the main hyphae, 33.3-50 µm in diameter and spherical, oogonial wall unpitted except where antheridial touch; antheridia monoclinous and diclinous, one to many per oogonium, attached laterally or apically. In some cases, antheridia are attached to the oogonium by projections; oospores 1-6 in number, commonly 3, 15-27 µm in diameter, centric.

Isolated from soil (pH 8.0), Government fish farm, Utrethiya, Lucknow (Misra, 1980); saprophytic on insects in Pune, September 1976 (Kapadnis, 1980), from alkaline ponds of Lucknow (Misra, 1982); from Brahma Sarovar, Kurukshetra (Gupta and Mehrotra, 1988a, 1989a & b).

Khare (1992) collected it from Jalalpur lake and Dhopeshwar pond of Bareilly. He observed abundant gemmae with variable shape (spherical, cylindrical, irregular), single or in chains, and oospores number from 2-22.

Achlya recurva Cornu

Amer. Sci. Nat. Bot. Soc. 5 : 15-22, 1872

(Plate 11; Fig. 12)

Hyphae coenocytic, up to 128 µm wide, gemmae present; zoosporangia clavate, 126-300 × 20-25 µm; oogonia abundant, lateral, spherical, 25.2-64.8 µm in diameter, covered with truncate wall ornamentation; oogonial stalk stout, recurved or bent; antheridial branches androgynous, clavate; oospores eccentric, 1-9 in number per oogonium, 20-32 µm in diameter, spherical to ellipsoidal.

It was isolated from Saugar (Saksena and Rajgopalan, 1958) scrub Jungle and dry water land soils, Hyderabad (Manoharachary, 1977); Tejpur and Bangalore (Chowdhry and

Agarwal, 1980b), Pune (Kapadnis, 1980), Hyderabad (Manoharchary, 1979b; Manoharchary and Rao, 1981), muddy soils of Mannanur forest, A.P. (Rao and Manoharachary, 1981) and two tanks of Nizamabad, A.P. (Bhairabnath and Manoharachary, 1985).

Achlya rodrigueziana Wolf

Mycologia *33* : 274, 1941

(Plate 12; Figs. 8-9)

Mycelium extensive, dense, sporangia branched; gemmae abundant, filiforms; sporangia abundant, filiform or fusiform, 150-670 × 20-87 μm, renewed sympodially; zoospores discharge achlyoid, spore cluster persistent at the mouth; encysted spore 9-10 μm in diameter; oogonia abundant, lateral, terminal, spherical, 25-31 μm in diameter; oogonial wall unpitted; antheridia present, diclinous; antheridial cells branched; fertilization tubes present; persistent; oospores eccentric, aplerotic, one in number, 22-28 μm in diameter (Verma, 1984).

This isolate was differentiated by monoclinous or androgynous antheridia, irregular inner wall of the oogonia, eccentric and single oospore in the oogonium (Verma, 1984).

This species was isolated on hemp seeds from pond waters near Manduadih, Varanasi on 12th September 1965, culture R.D. 9 (Dayal and Thakurji, 1969); water (pH 8.1) of Major's pond, Telibagh, Lucknow (Misra, 1980, 1982; and from cultivated soils of Nanakmatta, Nainital (Verma, 1984).

Achlya stellata de Bary

Current Science **44** (4) : 139

(Plate 13; Fig. 7)

Mycelium moderately extensive, diffuse; two week old colony 1.5-2.5 cm in diameter; principal hyphae slender, branched, 70-105 μm in diameter at base; zoosporangia moderately abundant, fusiform or clavate, less frequently naviculate or filiform, 275-550 μm long by 18-45 μm in diameter, predominantly 350-500 × 25-35 μm, renewed sympodially, occasionally cymosely; zoospores discharge achlyoid; spore cluster not persistent, encysted spores 7.3-10 μm in diameter, gemmae absent; oogonia abundant, terminal on main hyphae or on lateral branches which are commonly curved or rarely once coiled; oogonia spherical pyriform, often with a neck, the basal septum often having an inward projection; 35-84 μm in diameter, predominantly 55.5-70 μm inclusive of ornamentations; oogonial wall unpitted, densely ornamented with conical or mummiform papillae; antheridia usually absent or very sparse, present on about 1 per cent oogonia; when present normally one, monoclinous (androgynous) on a long or rarely short antheridial branch, antheridial cells thin walled, tubular or clavate; apically or very rarely laterally appressed to the oogonium; oospores usually one, rarely two in an oogonium, normally spherical but occasionally ovoid, filling the oogonium and maturing; oospores subcentric; 32.5-46.5 μm in diameter, predominantly 30-37 μm (Srivastava, 1975c).

It was collected from Jalwania Pond, Gorakhpur on January 12, 1974 (Srivastava, 1975c).

Achlya stellata de Bary var. *multispora* Rai and Misra

Current Science **46**(1): *28* : 1977

(Plate 12; Fig. 10)

Mycelium moderate, extensive, diffuse, hyphae slender, branched, 70-105 μm in diameter; gemmae absent; sporangia moderately abundant, fusiform or clavate, 155.5-550.0 × 18-45 μm; predominantly 350-500 × 25-35 μm; renewed sympodially, zoospore discharge achlyoid; encysted spores 5.54-8·3 μm in diameter; oogonia abundant on lateral branches, globose, measuring 35-88.6 μm in diameter, predominantly 55.5-70 μm; inclusive of ornamentation; oogonial wall unpitted; antheridia generally present, monoclinous, laterally appressed to the oogonium; oospores usually one, rarely two, (3-5, Rai and Mishra, 1977a) in an oogonium, spherical, filling the oogonium, 19.3-46 μm, mostly 30-37 μm in diameter, subcentric.

It was collected from alkaline muddy soil (pH 8.0), fisheries farm, Utrethiya, alkaline ponds, Lucknow (Rai and Misra, 1977a).

Plate - 9

Explanation of Plate -9

(Khulbe, 1985)

Figs. 1-3 : *Achlya americana* Humphrey.

1. Mature zoosporangia showing liberation of zoospores and sympodial branching.
2. Mature oogonia with monoclinous antheridia, eccentric oospores and conspicuous pits.
3. An oogonium with androgynous antheridium.

All Figs. scale A.

Fig. 4 : *Achlya flagellata* Coker.

Oogonia with diclinous antheridia and eccentric oospores. Scale A.

Figs. 5&6 : *Achlya klebsiana* Pieters.

5. Oogonia with monoclinous branches of antheridia enwraping the oogonia.
6. Mature oogonium with diclinous antheridium and eccentric oospores. Figs. 5 & 6, scale B

Figs. 7&9 : *Achlya orion* Coker & Couch.

7. Mature zoosporangium showing liberation of zoospores.
8. Young oogonia having curved stalks and monoclinous antheridia.
9. Mature oogonium with single eccentric oospore.

All Figs., scale B.

Figs. 10&11 : *Achlya prolifera* (Ness) de Bary.

10. Young oogonia.
11. Mature oogonium with long diclinous antheridium

Figs. 10 & 11, scale B.

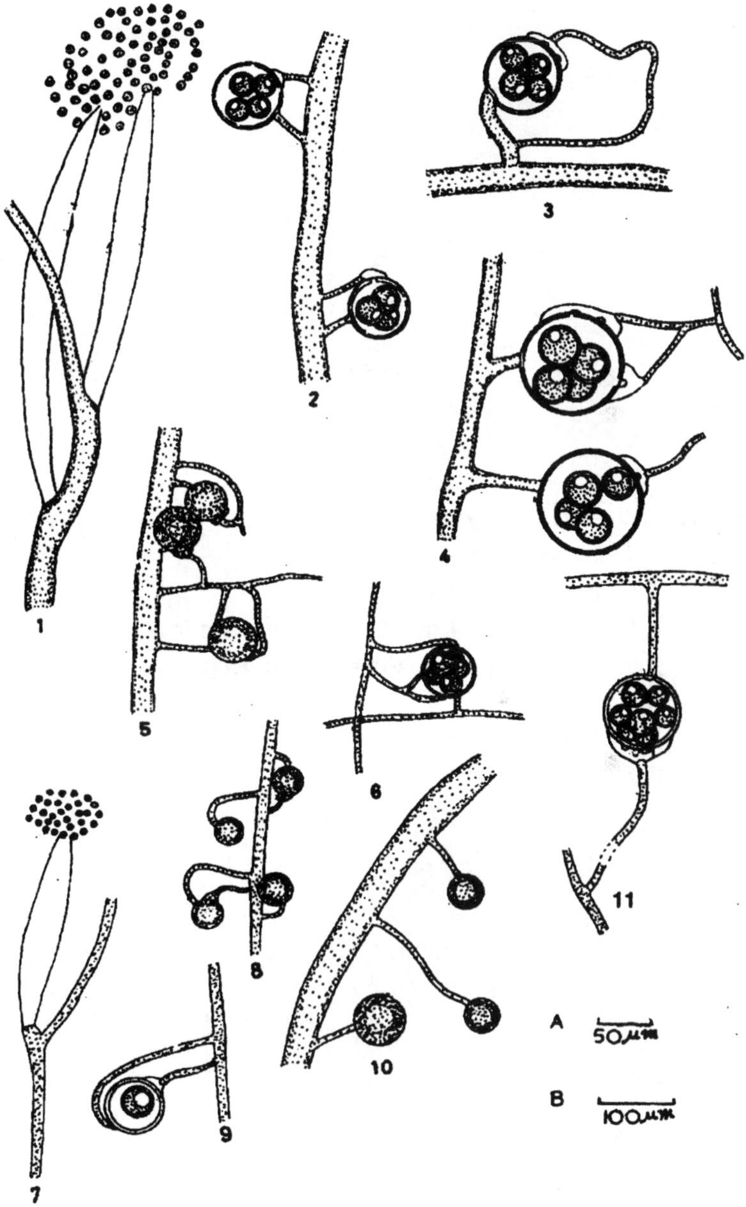

Plate 9

Plate - 10

Explanation of Plate - 8

(Khulbe, 1985)

Figs. 1-7 : *Achlya oryzae* Ito at Nagai.

1. Zoosporangia.
2-5. Various gemmae.
6. Lateral oogonium with diclinous antheridia.
7. Mature oogonium with eccentric oospores.

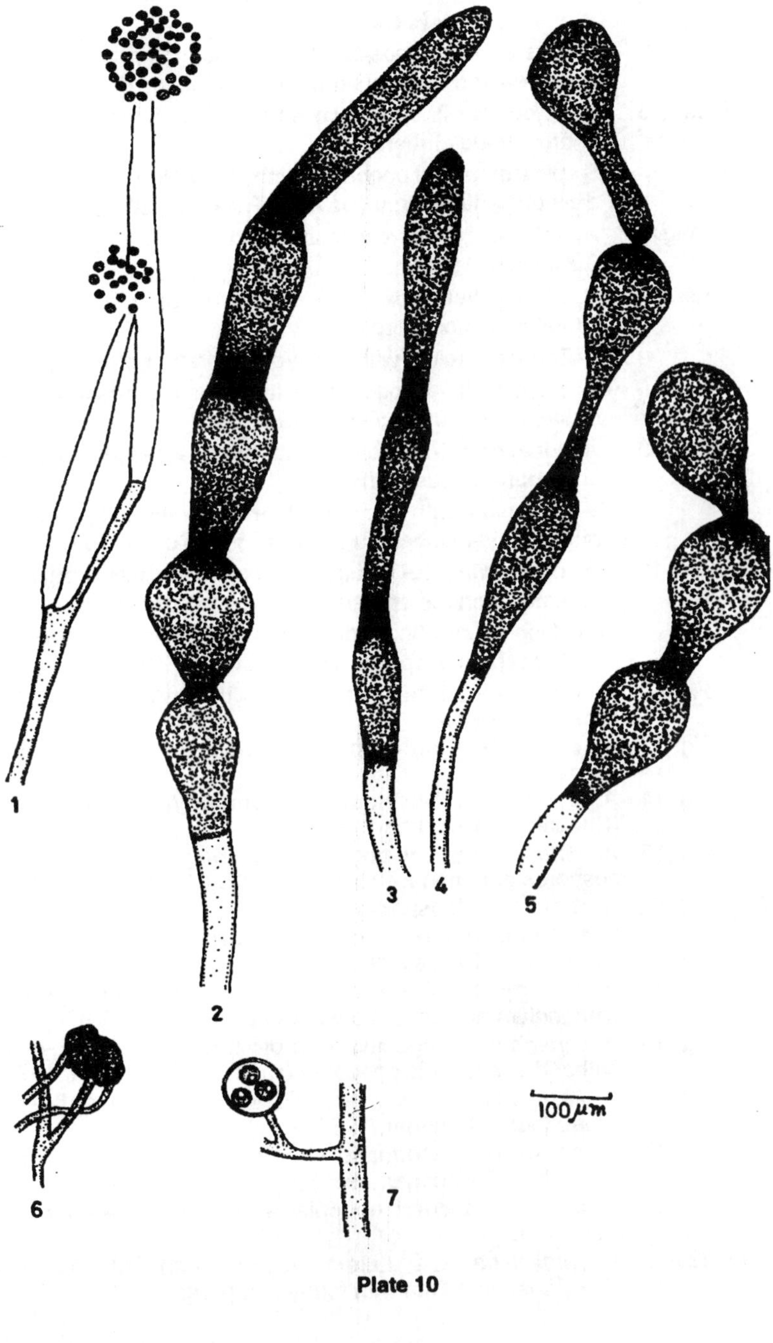

Plate 10

Plate - 11

Fig. 1. : *Achlya androcomposita* - Lateral oogonium with centric oospores and diclinous antheridia (Hamid, 1942).

Figs. 2-3 : *A. apiculata* : 2. oogonium with subcentric oospores and androgynous antheridium.

3-apical spherical oogonium with subcentric oospores and diclinous antheridium (Dayal & Thakurji, 1968a).

Fig. 4 : *A. aplanes*, zoospore with aplanoid zoospores (Chaudhuri & Kochhar, 1935).

Figs. 5-7 : *A. planes* var. *indica* : 5- zoosporangium with dictyoid liberation of zoospores.

6- Zoosporangium with achlyoid liberation of zoospores.

7- Short stalked oogonium with eccentric oospores and diclinous antheridia (Dayal, 1958).

Fig. 8-9 : *A. aquatica* : 8- Oogonium with many eccentric oospores and androgynous antheridia.

9- Oogonium with some papillae, eccentric oospores and monoclinous antheridium (Dayal & Thakurji, 1969).

Figs. 10-11 : *A. conspicua* 10- Lateral oogonium with androgynous and monoclinous antheridia and young oospores.

11- Oogonium with monoclinous and diclinous antheridia and eccentric oospores (Srivastava, 1967).

Fig. 12. *A. recurva* -Lateral oogonium with truncate wall and eccentric oospore.

Fig. 13. *A, crenulata*. Oogonium with papillate wall, eccentric oospores.

Fig. 14. *A. debaryana*. Diclinous and androgynous antheridia (Chaudhury & Kochhar, 1935).

Fig. 15. *A. diffusa* - Lateral oogonia with long stalks, eccentric oospores and monoclinous antheridia (Srivastava, 1967).

Figs.16&17a. *A. dubia*. 16- Zoosporangium,

17a- Oogonium with eccentric oospores and diclinous antheridium (Srivastava, 1967).

Fig. 17b. *A. hypogyna*. Pyriform oogonium with androgynous antheridium and centric oospores (Rai & Misra; 1976).

Fig. 18. *A. imperfecta* - Oogonium with diclinous and androgynous antheridia, eccentric oospores (Khulbe & Bhargava, 1977).

Fig. 19. *A. kashyapia* - Young oogonia with diclinous antheridia (Chaudhuri & Kochhar, 1935).

Fig. 20. *A. conspicua*. Oogonia with androgynous antheridia. (Chaudhuri & Kochhar, 1935).

Fig. 21. *A. cornuta* - Lateral oogonia with short stalks and no antheridia.

Fig. 22a&b. *A. caroliniana*. 22a- Lateral oogonia with 1-2 oospores, 22b- Zoosporangium with many exit pores.

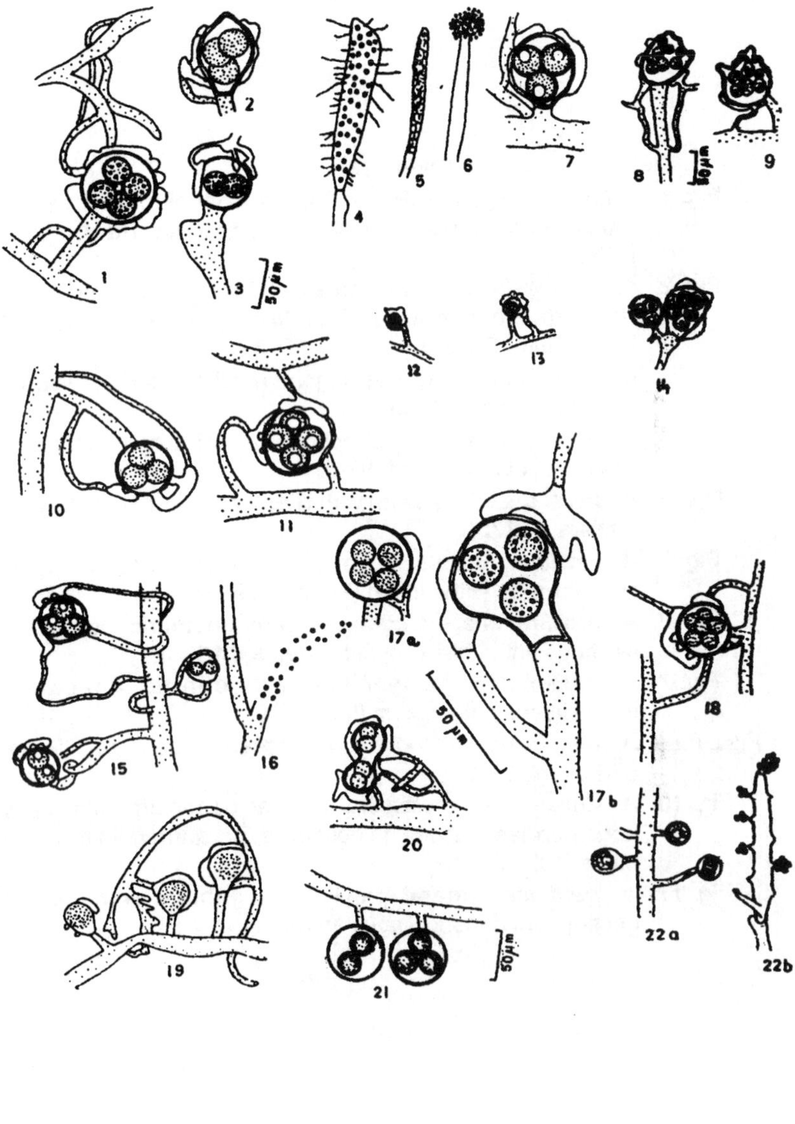
1
2
3
50 µm
4
5
6
7
8
50 µm
9
12
13
14
10
11
17a
15
16
50 µm
17b
18
20
19
21
50 µm
22a
22b

Plate 11

Plate - 12

Fig. 1. *Achlya oblongata* - Apical oblong oogonium with many subcentric oospores and diclinous antheridia (Rai & Misra, 1977b).

Fig. 2. *A. oblongata* var. *oblongata* - Oblong oogonium with subcentric oospores and diclinous, much branched antheridium (Dayal & Thakurji, 1968a).

Fig. 3a. *A. oligacantha* - Papillate oogonium with centric oospores (Dayal & Thakurji, 1965).

Fig. 3b. *A. megasperma* - Lateral oogonium with large subcentric oospores, diclinous antheridia.

Fig. 4. *A. polyandra* - Oogonium with many centric oospores (Rai & Misra, 1977b).

Fig. 5. *A. prolifera* - Lateral oogonium with eccentric oospores and diclinous antheridium (Khulbe, 1985).

Fig. 6. *A. proliferoides* - Oogonium with eccentric oospores, diclinous antheridia (Dayal & Thakurji, 1965).

Fig. 7. *A. racemosa* - Oogonium with centric oospores, monoclinous antheridium (Misra, 1980)

Figs. 8 &9 *A. rodrigueziana* - Oogonia with eccentric oospores (Dayal & Thakurji, 1969)

Fig. 10. *A. stellata* var. *multispora*. Lateral oogonium with many large papillae, subcentric oospores, no antheridium (Rai & Misra, 1977a).

Fig. 11. *A. papilosa*. Lateral papillate oogonium with centric oospores and monoclinous antheridium.

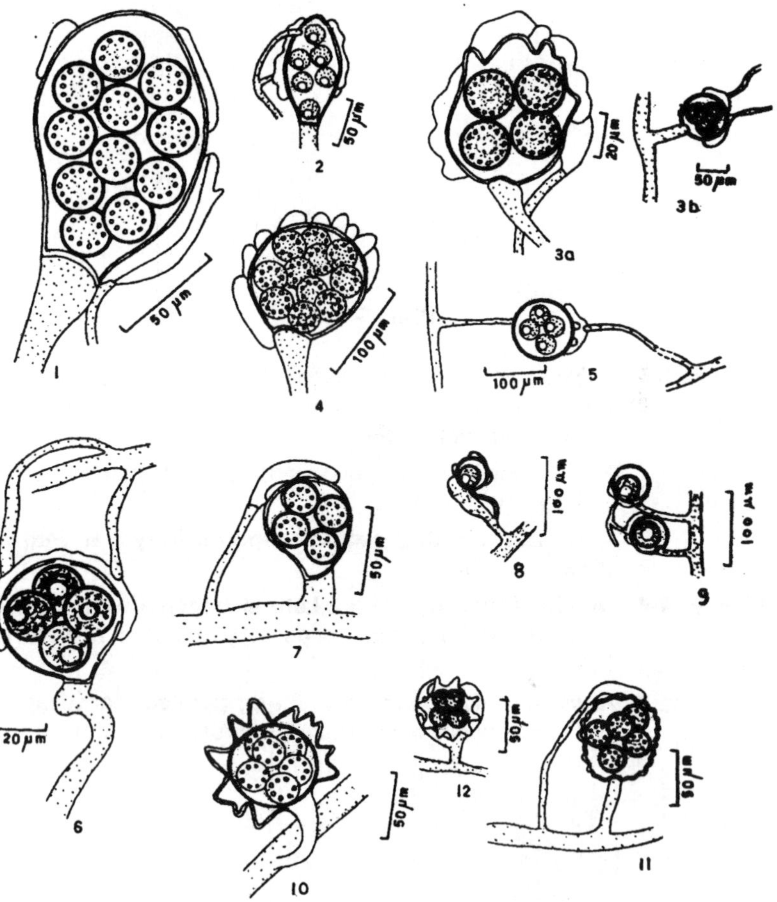
1
50 µm
2
50 µm
3a
20 µm
3b
50 µm
4
100 µm
5
100 µm
6
20 µm
7
50 µm
8
100 µm
9
100 µm
10
50 µm
11
50 µm
12
50 µm

Plate 12

Plate - 13

Figs. 1-4. *Achlya colorata* : 1. Oogonium showing outer smooth and inner irregular wall, centric oospores and attendant androgynous antheridia.

2 & 3- Oogonia with young oospores, monoclinous and androgynous antheridia.

4- Oogonium with 2 centric oospores (Dayal & Thakurji, 1966).

Figs. 5 & 6. *A. dubia* var. *pigmenta* : Lateral oogonia with diclinous antheridia and eccentric oospores (Chaudhuri & Kochhar, 1935).

Fig. 7. *A. stellata* : Oogonium with sharp, pointed many papillae and androgynous antheridium (Srivastava, 1975c).

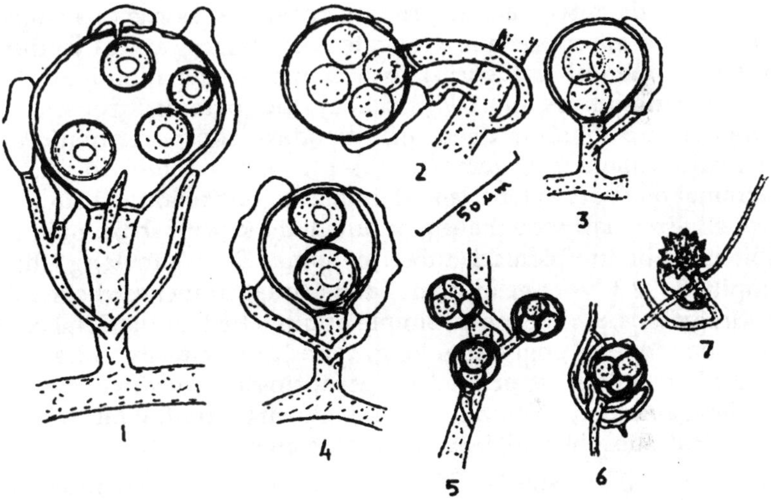

Plate 13

Aphanomyces de Bary

Jahrb. Wiss. Bot., 2 : 178, 1860

Aquatic Phycomycetes, p. 843, 1960

The Monograph of the Genus *Aphanomyces*, p. 1-95, 1961

Hyphae slender, delicate, long, sparingly to much branched, 3-11.5 µm in diameter, hyaline to light brown, regular or contorted, aseptate; zoosporangia filamentous of variable length, isodiametric, rarely tapering towards the apex, formed from unchanged vegetative hyphae, not proliferating within old zoosporangia, rarely lateral; primary zoospores borne in a single row in the zoosporangia, elongated, emerging apically, then become spherical and encyst in a clump at the mouth of the zoosporangium as in *Achlya* ; encysted primary zoospores spherical, 5.5-15 µm in diameter; secondary zoospores reniform, laterally biflagellate; gemmae absent; oogonia few to many, terminal on short to long lateral branches; oogonial wall smooth walled, irregularly roughened, or ornamented with sharp pointed spines or bluntly conical, tubercle; spines 2-3.5 µm long, thin, unpitted. 12 to 40 µm in diam.; antheridial branches diclinous, androgynous, monoclinous, simple or branched; antheridial cells one to many per oogonium, long, cylindrical, clavate, tuberous, laterally or apically appressed; oospores single, rarely two, hyaline or dark, aplerotic, eccentric, with a single large central oil globule, spherical, subspherical, 8-35 µm in diameter.

In India, 13 species of *Aphanomyces* have been found in different habitats till 1998. Their distinguishing characters are as follows :

Key to the species of *Aphanomyces*

Oogonia smooth walled .... A

Oogonia ornamented .... B

A_1 Outer and inner wall smooth C

A_2 Outer wall smooth, inner wall irregular .. *A. euteiches*, p.100

B_1 Antheridial branches diclinous .... F

B_2 Antheridial branches diclinous and .... monoclinous both, oogonial wall with curved spines, ... G

C_1 Oogonia few or rare .... D

C_2 Oogonia common, abundant .... E

D_1 Antheridia mostly diclinous; rarely androgynous, ... *A. laevis*, p. 102

D_2 Antheridia mostly monoclinous or/and androgynous, rarely diclinous .... *A. keratinophilous*, p. 101

E_1 Antheridia strictly diclinous ....*A. patersonii*, p. 103

E_2 Antheridia diclinous or monoclinous, forming helicoidal, spiral about the oogonial stalk ....*A. helicoides*, p.100

E_3 Antheridia develop near oogonial base on short stalk .... *A. brassicae*, p. 98

F_1 Oogonial wall with curved spines, oospores golden brown .... *A. sparrowii*, p.104

F_2 Oogonial wall - outer smooth and sinous inner contour, oospores hyaline to yellow..*A. cochlioides*, p. 99

G_1 Oogonial wall with bluntly conical tubercle, oospores one to two .... *A. stellatus*, p. 105

G_2 Oogonial wall with many, short pointed spines, oospores hyaline .... *A. scaber*, p. 104

G_3 oogonial wall with short, blunt, cone shaped projections, oospores brown .... *A. apophysii*, p.98

G_4 Oogonial wall irregularly thick, but no spines or tubercle .... H

H_1 Antheridial branches long, oospore hyaline.. *A. cladogamous*, p. 99

H_2 Antheridial branches short, oospores hyaline or light brown .... *A. irregulare*, p. 103

Aphanomyces apophysii Lacy

Indian Phytopathology 2 : 137, 1949

(Plate 14; Figs. 18-2C)

Hyphae 8 µm in diameter, coarse, hyaline, irregularly branched, with occassionally regular swellings (apophyses), 20-30 µm in diameter, forming a contorted intramatrical system; zoosporangia long, filamentous, formed from intramatrical hyphae which penetrate the lateral walls of the host cells, 8-12 µm in diameter, swollen basally, tapering apically, primary zoospores few in number, elongate with rounded ends, encysting upon emergence at the orifice; primary zoospore cysts 10-12 µm in diameter, contents emerging through an irregular aperture; secondary zoospores not observed; oogonia terminal on short, lateral branches, intramatrical, spherical, 35-40 µm in diameter with a thick wall ornamented with short, bluntly one-shaped projections up to 5 µm in length; oospores 20-24 µm in diameter, brown contents coarsely granular; antheridia one to several, small, clavate, antheridial stalks unbranched, diclinous or monoclinous in origin; germination of oospores by the formation of a germ tube.

Lacy (1949) found it as a parasite of *Spirogyra* sp. This species was characterized by the presence of conspicuous swellings in the intramatrical mycelium, most noticiable at points where the hyphae penetrate the transverse walls of the algal cells, and similar enlargement occurs at the base of the zoosporangia. This species also exhibited tendency to attack first the male filaments of the conjugating *Spirogyra*, and later the female filament and zygospores (Scott, 1961).

Aphanomyces brassicae Pavgi and Singh

Mycopathologia *61* (3) : 167-172, 1977.

(Plate, 14; Figs. 11-14)

Hyphae hyaline, delicate, 4.5-8.5 µm in width, moderately branched approximately at right angles; zoosporangia terminal, moderately long, 4.5-7.5 µm in width, formed from the undifferentiated vegetative hyphae, never tapering towards the apex, of even diameter from basal to distal end; primary zoospores spherical, 20-30 in number, encysting upon emergence above the orifice; primary zoospores cysts, 5-8.5 µm in diam.; secondary zoospores ovate to reniform, laterally biflagellate; oogonia terminal on short branches, primary subglobose, thin walled measuring 24.5-35.5 µm in diam.; antheridia arising from a simple, stout, clavate stalk, measuring 5.5 - 8.5 × 8-14.5 µm, always single, diclinous; oospores spherical, thick walled with smooth outer wall, hyaline, measuring 16.5-21.5 µm in diam., germination either by formation of germ sporangium or by direct production of vegetative mycelium and zoosporangia (Singh and Pavgi, 1977).

It was collected as a parasite of living roots of *Brassica oleracea* L. var. *botrytis* at Varanasi, U.P. on 23rd September, 1971, leg. S.L. Singh (Type) (I.M.I-188398) (Singh and Pavgi, 1977).

Aphanomyces cladogamous Drechsler

J. Agric, Res. *38* : 335, 1929

(Plate 14; Figs. 4-7)

Hyphae 3.4-6.8 µm in diameter, delicate, hyaline; zoosporangia very large often extensive involving the conversion of large segments of vegetative hyphae, not tapering towards the apex, zoospores encysting at the orifice upon emergence; oogonia terminal on short lateral branches of variable length, subspherical to globose, 40.8 µm in diameter, mostly averaging 28.9 µm; walls irregularly thickened with a smooth outer surface; oospore single, hyaline, 13.6-23.8 µm in diameter, thick walled; contents regular and with a small conspicuous refractive body in the centre; antheridia 2-3, consisting of an inflated portion of variable length antheridial stalk long, diclinous or monoclinous in origin; germination of the oospores could not be seen (Dayal and Thakurji, 1965).

It was isolated from ditch water, Agriculture farm, Varanasi (Dayal and Thakurji, 1965); and soil samples of Lucknow and Bangalore (Chowdhry and Aggarwal, 1980b).

Aphanomyces cochlioides Drechsler

Phytopathology, **18**: 149, 1928

J. Agric. Res., **28**: 326, 1929

A Monograph of the Genus *Aphanomyces*, p. 38, 1961

(Plate 14; Fig. 17)

Hyphae 3-10 µm in diameter, hyaline, slender, branched; zoosporangia long, irregular in thickness, tapering towards the apex, with small branches; primary zoospores elongate, 200-350 per zoosporangium, encyst at the mouth of the zoosporangia, encysted primary zoospores 6-16 µm in diameter, secondary zoospores reniform, laterally biflagellate; oogonia lateral on small stalks, subspherical, 20-29 µm in diameter, oogonial wall irregularly thickened, outer wall smooth, inner wall shows sinuous; antheridial branches diclinous, branched, 1-5, curved, clavate; oospore one, 15-25 µm in diameter, hyaline to yellow, eccentric with a large central oil globule.

Chowdhry and Aggarwal (1980 a & b) isolated it from soil and water samples around IARI, New Delhi during winter when temperature ranged between 10-25°C. They recorded it for the first time.

Aphanomyces euteiches Drechsler

J. Agric. Res., **30**: 311-312, 1925

A Monograph of the genus *Aphanomyces*, p. 40-41, 1961

(Plate 14; Fig. 16)

Hyphae 3.5-11 µm in diameter, delicate, hyaline, branched; zoosporangia many, long, branched, tapering; primary zoospores elongate, 4-50 µm in length; often 300 or more in number, encyst at the tip of zoosporangium; encysted zoospores 8-12 µm in diameter; secondary zoospores reniform, laterally biflagellate; oogonia lateral on short branches, spherical, subspherical, 22-36 µm in diameter, thick walled with smooth outer wall and sinuous inner wall; antheridial branches diclinous, branched; antheridial cells clavate, curved, 1-5 per oogonium; oospore single, hyaline, 16-28 µm in diameter, thick walled, eccentric with a large central oil globule.

Chona, Lall and Kakariya (1958) in a survey of fungi of Delhi reported it on *Lathyrus sativus* from New Delhi (Dayal and Kiran, 1985).

Aphanomyces helicoides Minden

Kryptogamenfl. Mark Brandenburg, 5 : 559, 1915

(Plate 15; Figs. 1-6)

Hyphae delicate, hyaline, branched, 4.5-11.5 µm in diameter at base; zoosporangia filamentous, long, formed from undifferentiated vegetative hyphae; primary zoospores rod shaped or fusiform, encysted at the mouth of zoosporangium, encysted primary zoospores 8-11.5 µm in diameter, diplanetic; oogonia terminal on lateral branches of variable length, some in dense clusters, spherical, smooth-walled, 18-37.5 µm in diameter, mostly 30 µm; antheridia 1-4 per oogonium, large, cylindrical, closely wrapped about the oogonium, 5-7 µm thick; antheridial stalks simple or branched, diclinous or monoclinous, forming helicoidal spirals about the oogonial stalk, sometimes wrapping about themselves and around other adjacent hyphae; oospore single, spherical, aplerotic, eccentric, 16-35 µm in diameter, mostly 28 µm (Khulbe, 1985).

This species was reported by Srivastava (1967a) from water samples of Gorakhpur for the first time. Later, Prabhuji (1984) isolated it from soils of Gorakhpur. The isolate showed helicoidal nature of antheridial stalks and cluster of oogonia, but hyphal knots are absent. The oogonia and oospores are smaller than those of Scott (1961). This species was found to be a new parasite of *Tor tor* Ham. (Khulbe, 1985).

Aphanomyces keratinophilous (Ook. and Kob.) Seymour and Johnson

Mycologia **65** (6) : 1312-1318, 1973.

(Plate 15; Fig. 8)

Hyphae slender, branched, 4-8.5 µm in diameter, mostly 8.0 µm; zoosporangia filamentous, 300-750 µm long and 3.5-8.5 µm in diameter; zoospores develop in a single row, encyst in a spherical mass at the mouth of the zoosporangium; encysted zoospores 8.5 µm in diameter; oogonia rare, spherical, smoothwalled, unpitted, 12-23.5 µm in diameter, mostly 18.5 µm, lateral on very short branches; antheridial branches monoclinous, androgynous, rarely diclinous; antheridial cell clavate; oospore spherical, eccentric, single, 8-18 µm in diameter, mostly 16.5 µm (Khulbe, 1985).

Isolated on snake skin and human nails from cultivated soil of Dwarahat on February 10, 1982, leg. R.D. Khulbe. The type culture has been deposited in the Department of Botany, Kumaun

University, Nainital (Aph/2). This species was also collected from Khurpa Tal and Nainital lakes (Khulbe, 1985).

Aphanomyces laevis de Bary

Jahrb. Wiss. Bot., **2** : 179, 1860.

(Plate 15; Fig. 7)

Main hyphae slender, branched, 4-8 µm thick at base; zoosporangia filamentous, 350-800 µm long by 4-8.5 µm in diameter; zoospores arranged in a single row, rod shaped; encysted spores spherical, 5.5-8.5 µm thick in diameter, mostly 8.5 µm; oogonia few, terminal on short lateral branches, spherical or subspherical, 15-25.5 µm in diameter, predominantly 27.5 µm; antheridial branches numerous, diclinous, rarely androgynous, rarely wrapping around the oogonial branch of oogonium, 1-3 per oogonium; antheridial cells small, tuberous; oospore spherical, eccentric, single, 12-20 µm in diameter, predominantly 18.5 µm (Khulbe, 1985).

This species was isolated from Pusa (Butler, 1911), soil and water samples of Delhi, Rampur and Bangalore (Chowdhry and Aggarwal, 1980b); permanent, temporary and cemented ponds of Varanasi (pH 7-8.5) (Dayal and Thakurji, 1966); soils of Jyolikot and Naina peak (Sati and Khulbe, 1980); on hemp seeds and snake skins in Poona (Kapadnis, 1980); lake waters of Nainital and Sariya Tal (in Summer & Spring), Khurpa Tal, Bhim Tal and Naukuchia Tal (in all the seasons) at temperature 10-28.5°C (Khulbe, 1977; Khulbe and Bhargava, 1977) and 7.3-8.8 pH (Khulbe, 1980a); Ram Tal, Nal Damyanti Tal and Sat Tal during summer and rainy seasons at 22-25°C temperature (Mer *et al.*, 1980; Mer and Khulbe, 1984); ponds A&B, Hyderabad during all the seasons (Manoharachary and Rao, 1981); from Deotal pond, Jabalpur during winter, summer and monsoons (Agarwal *et al.*, 1989); on hemp seeds and snake skins from surface waters, decaying plant materials from Ganga Sagar Jabalpur during November to March and in April (1987-1988) at 16–26.5°C temperature, 7-7.5 pH and excess nutrients (Mishra *et al.*, 1990); on snake skins from soils of Kilbury, Naini Tal and different cultivated and forest soils, lakes, ponds, tank streams and rivers of Kumaun Himalaya at 5-30°C temperature, 2-90 per cent moisture, 5.3-6.9 pH and 0.4-4.9 organic carbon (Khulbe, 1985 & 1991).

This species was also isolated on snake skin bait from Brahma Sarovar and Sanhit Sarovar, Kurukshetra (Gupta and Mehrotra, 1988). In the recent past, Khare (1992) found it as a frequent water mold in river, pond and lake waters of Bareilly.

Aphanomyces irregulare Scott

A Monograph of the genus *Aphanomyces*, p. 47, 1961

(Plate 14; Figs. 1-3)

Hyphae delicate, hyaline, sparingly branched; zoosporangia long, filamentous, 5.5-7.4 μm in diameter, formed from undifferentiated hyphae; isodiametric; zoospores encysting upon emergence at the orifice, 7-11 μm in diameter; oogonia terminal on short lateral branches, spherical, 18.5-30 μm in diameter; wall roughened due to irregularities but never with definite spines or tubercle; oospores 16-22 μm, hyaline or light brown, content finally granular with a large central oil globule; antheridia single, rarely two, small, clavate, antheridial stalk unbranched, short, diclinous or monoclinous; fertilization tubes not observed; germination of oospore by formation of a germ tube (Dayal and Thakurji, 1968b).

It was isolated on hemp seed from pond water, behind B.H.U. Press, Varanasi by Thakurji on 6 Nov. 1966. Slide R.D. 19 (Dayal and Thakurji, 1968b).

Aphanomyces patersonii Scott

The Virginia J. of Science 7 : 171, 1956

(Plate 14; Figs. 8-10)

Hyphae up to 7.2 μm in diameter, intramatrical, sparingly branched, ramifying throughout the body cavity of the host; zoosporangia long, filamentous, unbranched, of the same diameter as the vegetative hyphae, isodiametric, penetrating the host wall and extending at right angle into the surroundings medium; zoospores rod shaped, few in number, encysting at the orifice, or swimming by means of two forwardly directed flagella; encysted primary zoospores 10-15 μm in diameter; discharge schistose, secondary zoospores reniform; oogonia terminal on short lateral branches, sphericai, 23.8-37.4 μm in diameter, smooth walled; oospore 20.4-30.6 μm in diameter, hyaline, content granular with a single large oil globule; antheridia single, swollen clavate;

antheridial stalk simple, unbranched, diclinous in origin; oospore germination not observed (Dayal and Thakurji, 1965).

Isolated from water and soil samples of Varanasi (Dayal and Thakurji, 1965; Thakurji, 1967).

Aphanomyces scaber de Bary

Jahrb. Wiss. Bot., 2 : 178, 1860

A Monograph of the genus *Aphanomyces*, p. 53,54, 1961

(Plate 14; Fig. 23)

Hyphae 4-8 µm in diameter, slender, less branched; zoosporangia long, filamentous, entire vegetative hyphae convert into zoosporangium; primary zoospores elongated, becoming spherical upon encystment at the tip of the zoosporangium, encysted zoospores 8-9.5 µm in diameter; secondary zoospores reniform, laterally biflagellate; oogonia lateral on short branches, 12-20 µm in diam.; oogonial wall ornamented with many short, blunt or short pointed spines up to 3.5 µm in length; antheridial filaments diclinous, rarely androgynous, unbranched; oospore single, hyaline,10-18 µm in diameter, eccentric with a large central oil globule.

The species was isolated on hemp seeds from lake waters of Nainital (during summer), Khurpatal (summer to rainy seasons) and from Bhimtal and Naukuchiya tal (in all the seasons except autumn and winter) when temperature ranged from 15-28.5°C (Khulbe and Bhargava, 1977) and pH between 7.5-8.2 (Khulbe, 1980b). Kiran and Dayal (1983) isolated it on snake skin from Maidagin pond water, Varanasi (Slide No. 37). Gupta and Mehrotra (1988a) also reported it from Sanhit Sarovar of Kurukshetra.

Aphanomyces sparrowii Cutter

Mycologia 33 : 233, 1941

A Monograph of the genus *Aphanomyces*, p. 57, 1961

(Plate 14; Figs. 21-22)

Hyphae 6-15 µm in diameter, intramatrical, hyaline, sparingly branched; zoosporangia short, unbranched, lateral branches of the vegetative hyphae turn into zoosporangia, tapering; primary encysted zoospores 7-10 µm in diameter, few in number; oogonia lateral on short branches, spherical, 15-26 µm in diameter; oogonial wall with sharp, pointed spines up to 15 µm in length; antheridial filaments diclinous, unbranched; antheridial cells 1-3 per oogonium, clavate; oospore single, golden brown, 13-21 µm in diameter.

Kapadnis (1980) reported it as a saprophyte on snake skin from Pune in September, 1977.

Aphanomyces stellatus de Bary

Jahrb. Wiss. Bot. 2 : 178, 1860

Geobios **2** : 192, 1975.

(Plate 14; Fig. 15)

Hyphae delicate, hyaline, sparingly branched, 4-7.5 µm in diameter; zoosporangia filamentous, unbranched, formed from undifferentiated vegetative hyphae, 4-5.5 µm in diameter; primary zoospores elongate with rounded ends, encysting at the orifices upon emergence; primary zoospores cysts 7-10 µm in diameter; cluster of 70-100 cysts formed at the orifices; secondary zoospore reniform, laterally biflagillate; oogonia terminal on lateral branches of variable length, spherical, 23.5-38.5 µm in diameter; wall ornamented with bluntly conical tubercles, sometimes exhibiting short dichotomising, projections up to 6.5 µm in length; antheridia one to several, tubular or clavate, apical or laterally appressed to oogonium; antheridial branches diclinous or monoclinous, short to long, usually unbranched, sometimes slightly coiled; oospore one, rarely two, 15-28.5 µm in diameter, contents finely granular with a conspicuous central oil globule (Srivastava, 1975b).

It was collected on hemp seed from Laldiggi pond, Gorakhpur on January 3, 1974 by G.C. Srivastava for the first time (Srivastava, 1975b).

Later, this species was reported as a saprophyte from Pune in September, 1978 (Kapadnis, 1980). It was also found as a keratinophilic by Hasija and Khan (1983 & 1987) who isolated it from decaying human nail in fresh water pond (temperature 21-23°C; pH 8.0) of Engineering College and Deotal lake of Jabalpur (M.P.) during December 1979 to March 1980 (Specimen HCIO, No 33743).

Plate—14

Figs. 1-3. *Aphanomyces irregulare.* 1. zoosporangium, 2&3-oogonia with irregular wall, diclinous or monoclinous antheridia and eccentric single oospore (Dayal & Thakurji, 1968b).

Figs. 4-7. *A. cladogamous.* 4- filamentous zoosporangium .with ball of encysted zoospores, 5&7-lateral oogonium with diclinous antheridium and mature eccentric & plerotic oospore, 6-monoclinous antheridium with young aplerotic oospore. (Dayal & Thakurji, 1966).

Figs. 8-10. *A. petersonii* 8&9-oogonia with androgynous antheridia 10-apiculate oogonium with androgynous and diclinous antheridia (Dayal & Thakurji, 1965).

Figs.11-14 *A. brassicae.* 11-sporangium with irregular colony of encysted zoospores, 12&13-young oogonia and antheridia. 14-mature oogonium with eccentric oospore (Singh & Pavgi, 1977).

Fig. 15. *A. stellatus* -oogonium with many tubercles and monoclinous antheridium (Srivastava, 1975).

Fig. 16. *A. euteiches* -oogonium with inner irregular wall and diclinous antheridia

Fig. 17. *A. cochlioides* -oogonium with coiled antheridial branch (Chowdhry & Aggarwal, 1980 a&b).

Figs.18-20. *A. apophysii.* 18-parasitic hyphae with sporangia,

19-20-oogonia with papillae & diclinous antheridia

Figs.21&22. *A. sparrowii.* 21-small zoosporangium.

22-oogonium with sharp, long papillae and diclinous tuberous antheridium..

Fig. 23. *A scaber.* oogonium with small spines and diclinous antheridium (Khulbe, 1980b).

Scales. a-figs. 1-3, b-figs. 4-6, c-figs. 8-10, 18-20 and 21-22.

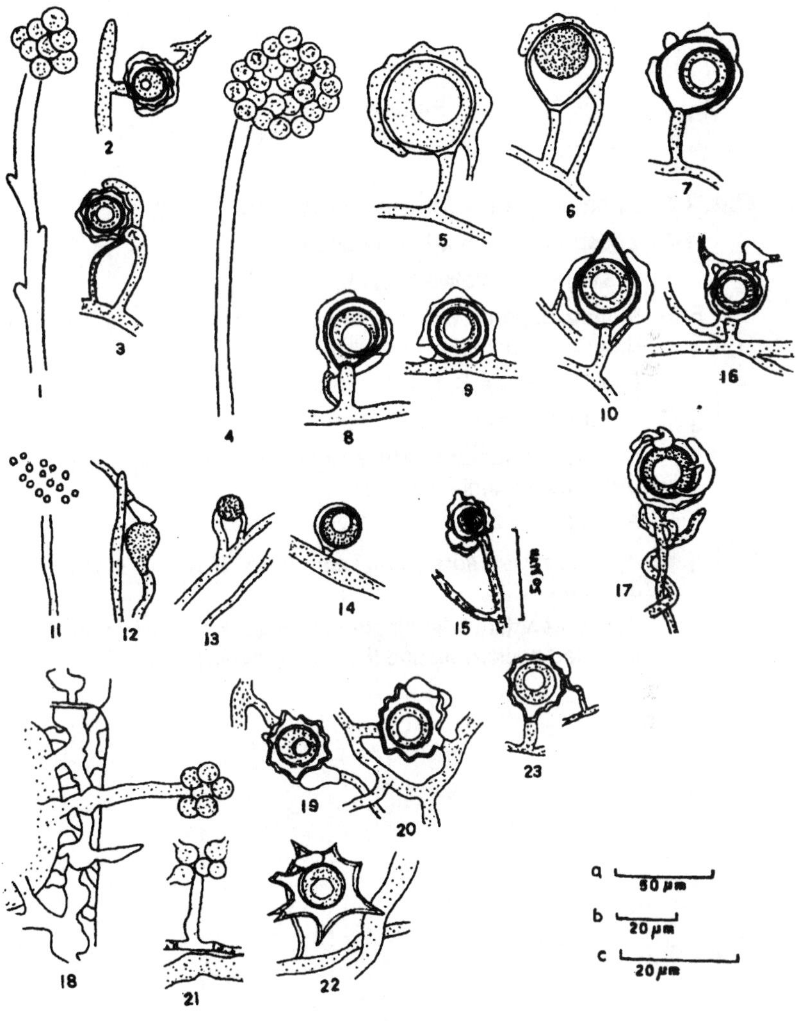
1
2
3
4
5
6
7
8
9
10
16
11
12
13
14
15
50 µm
17
18
19
20
23
21
22
a 50 µm
b 20 µm
c 20 µm

Plate 14

Plate-15

(Khulbe, 1985)

Figs. 1-6. *Aphanomyces helicoides* von Minden.

1-3. Zoosporangia showing zoospores.

4. Coiling of vegetative hyphae.

5&6. Mature oogonia with single eccentric oospore and coiled diclinous branches of antheridia.

Figs. 1-5, scale B; Fig. 6, scale A.

Fig. 7: *Aphanomyces laevis* de Bary.

Mature oogonium with single eccentric oospore and diclinous antheridial branch

Scale C.

Fig. 8: *Aphanomyces keratinophilous* (Ook & Kob.) Seymour & Johnson.

Zoosporangium with single row of zoospores and partially liberated zoospores and their encystment. Scale C.

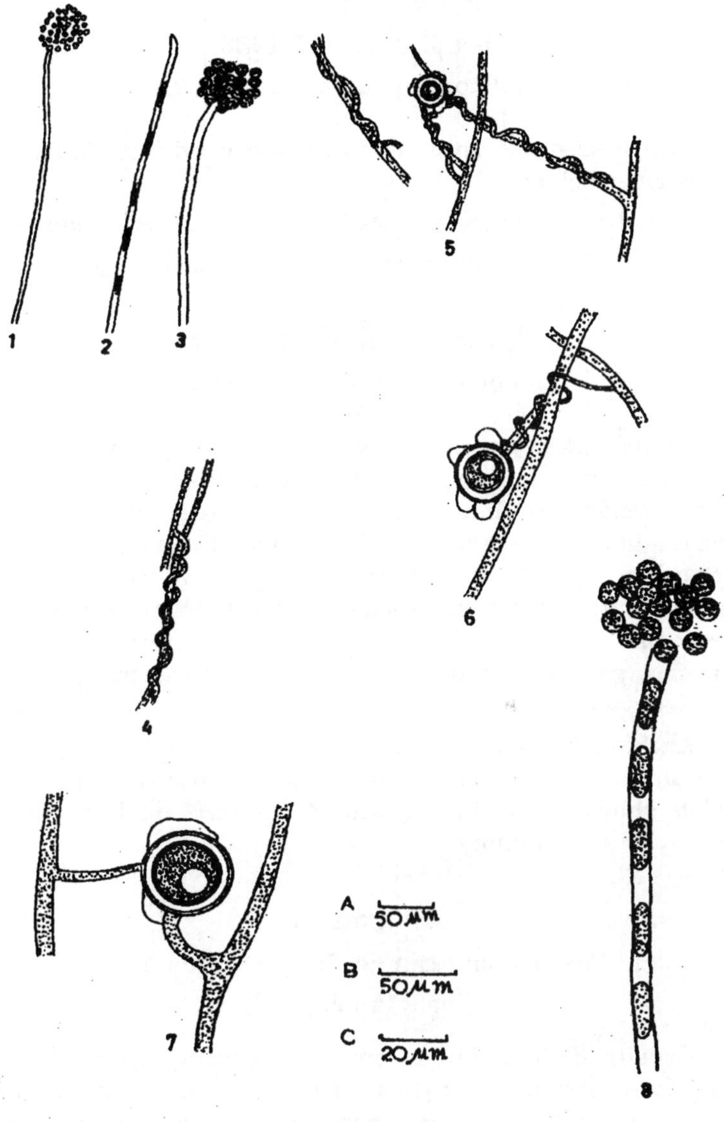

Plate 15

Aplanes de Bary

Bot. Zeit. **46** : 613, 1988

The Saprolegniaceae, p. 76, 1923

Two species of *Aplanes* have been reported from India. Their key is given below :

Sexual reproduction observed .... *A. braunii*, p. 110

.................not observed .... *A. sp.*, p.110

Aplanes braunii de Bary

Bot. Zeitung, **46** : 650, 1888

Hyphae 24-53 µm in diameter; zoosporangia terminal, very sparse, cylindrical; zoospores free within them, but irregularly many seriate, at times escaping, at times retained in the sporangium and sprouting there; oogonia rare, terminal or intercalary, spindle shaped, barrel shaped or pyriform, when terminal often with a snout-like elongated tip, wall heavy with many pits, oogonia 365 µm long by 88 µm wide; antheridial branches generally androgynous, short, unbranched; oospores not observed

Khare (1992) isolated this species from Jalalpur lake, Bareilly two times, but could not raise its culture in any medium. His isolate showed resemblance with *A. braunii* de Bary, Syn. *A. androgynous* Humphrey.

Aplanes sp.

Proc. Indian Acad. Sci. II (2): 137-154, 1935

(Plate 14 : Figs 1-7)

Growth delicate but extensive, reaching up to 2 cm on coagulated egg yellow; hyphae slender, occasionally branched, 9.2-15.3 µm in thickness; zoosporangia very rare, when present cylindrical to long clavate, 28-30.3 µm, renewed by internal proliferation, one zoosporangium observed with cymose branching; chlamydospores abundant, very variable in size and shape, elliptical to globular or irregularly cylindrical, terminal or intercalary, in chains or single, sexuality not observed.

It was isolated from Lahore, Amritsar and Gujranwala for the first time from India (Chaudhuri and Kochhar, 1935).

Aplanopsis Hohnk

Veroff. Inst. Meer. Brem. 1:85, 1952

Aplanopsis spinosa Dick

Trans. Brit. Mycol. Soc. **43**(1): 60-64, 1960

(Plate 14; Fig . 8)

Hyphae 40-82 µm in diameter, branching frequently, lateral hyphae arising almost at right angles to the main hyphae, sparingly branched; zoosporangia and gemmae absent; oogonial branches produced laterally on distal parts of some principal hyphae; oogonia abundant, formed laterally or terminally as many of the oogonial branches; spherical or subspherical, densely covered with papillate or tuberculate, curved or hooked projections, diameter range of oogonia 14.5-33 µm; antheridia androgynous or semi-hypogynous; oospores spherical or irregular, usually single, subcentric, 14-31.5 µm in diameter; germination of oospores not observed.

This species was isolated by Khare (1992) from Jalalpur lake, Bareilly.

Plate-16

Figs. 1-7: *Aplanes* sp.

1,2,4,&5- Hyphae showing young zoosporangia.

3- Empty zoosporangium with internal proliferation.

6 & 7- Gemmae behave as zoosporangia (Chaudhuri & Kochhar, 1935).

Fig. 8- *Aplanopsis spinosa* -A mature oogonium with spiny wall, single oospore, androgynous antheridium (Khare, 1992).

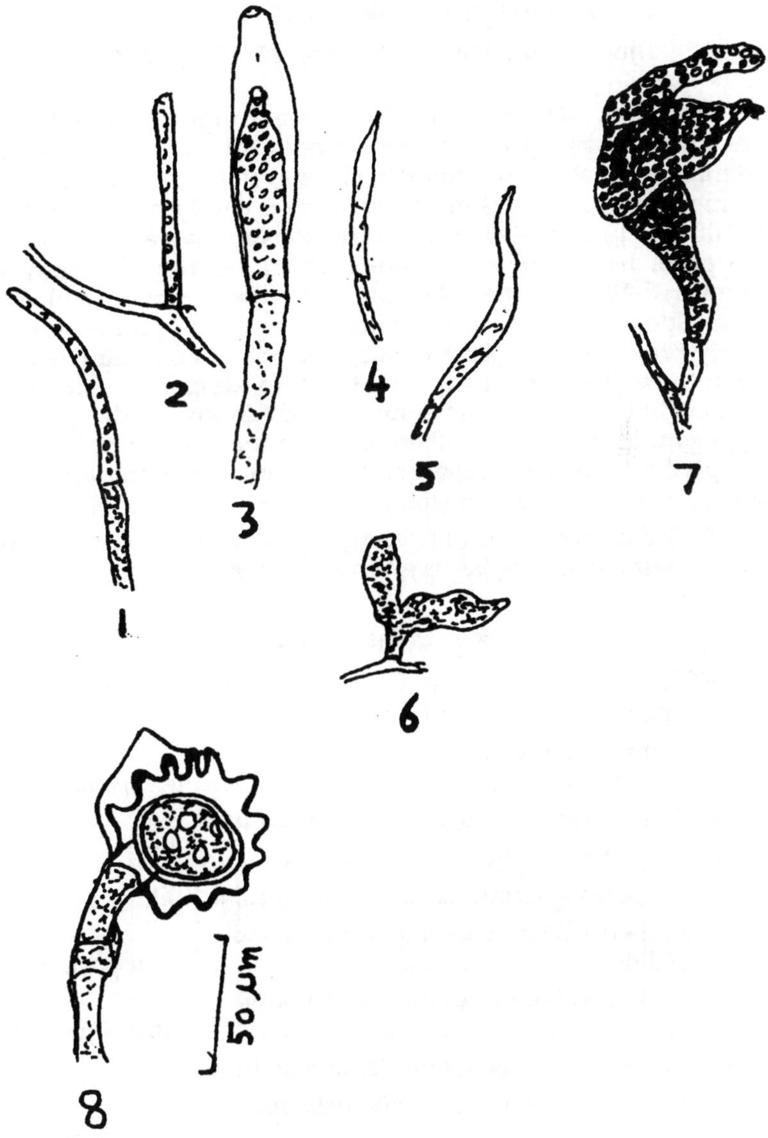

Plate 16

Brevilegnia Coker and Couch

Jour. Elisha Mitchell. Sci. Soc. **42** : 212, 1927

Mycelium slender, dense; hyphae sparingly branched, 4-65 µm thick at base, primary zoosporangia terminal, occasionally lateral, few to abundant, filiform, clavate, cylindrical, naviculate, filamentous, sometimes in cluster, simple, rarely branched, 8-35 × 100-700 µm in size; secondary zoosporangia sympodial, basipetal, lateral; zoospores arrenged in 1-5 rows, 1-many in number; 5-10 µm in diameter; zoospores liberate by dissolution of zoosporangial wall, some germinate in situ; gemmae absent, rarely present in *B. megasperma*; oogonia few to abundant, lateral, smooth walled, spherical, 18-48 µm in diameter, pits few, antheridial branched diclinous, monoclinous, androgynous, hypogynous, or absent, antheridial cell short, tuberous, laterally or apically appressed; oospore single, spherical, plerotic, aplerotic, eccentric, 15-50 µm in diameter.

In India, five species of *Brevilegnia* have been reported till 1998. Their identifying key is given as follows :

Key to the species

Zoospores strictly in single row ______ *B. linearis*, p. 117

Zoospores in one to many rows ______ A

A_1 Gemmae abundant, oogonia usually absent .... *B megasperma*, p.118

A_2 Gemmae absent, oogonia usually present B

B_1 Oospores plerotic C

B_2 Oospores aplerotic .... D

C_1 Antheridial branches monoclinous and diclinous .... *B. bispora*, p. 115

D_1 Antheridial branches mostly diclinous .... *B. diclina*, p. 115

D_2 Antheridial branches mostly monoclinous E

E_1 Antheridial branches rarely diclinous, often lacking .... *B. subclavata*, p.129

E_2 Antheridial branches rarely androgynous, oogonia sometimes irregular .... *B. irregularis*, p.117

E_3 Antheridial branches monoclinous, diclinous as well as androgynous .... *B. indica*, p. 116

Brevilegnia bispora Couch

J. Elisha Mitchell Sci. Soc. **42** : 227-242, 1927

(Plate 17; Figs. 1-4)

Mycelium dense, hyphae 54-58 μm in diameter at base; gemmae absent; sporangia abundant, filiform or clavate, renewed sympodially; 100-180 × 25-35 μm in diameter; spores discharge brevilegnoid; encysted spores 8-10 μm in diameters; oogonia abundant, lateral, spherical to subspherical with thick unpitted smooth wall and 19-26 μm in diameter; oogonial stalk slightly twisted in some cases; antheridial branches sparse, long., monoclinous, diclinous; antheridial cells lobed and laterally appressed; oospore single, spherical, plerotic, 16-19 μm in diameter, eccentric.

This species was isolated from muddy soils (pH 7.5) at Lacchi Tara, Telibagh, Lucknow (Misra, 1980), alkaline pond of Lucknow (Misra, 1982).

Brevilegnia diclina Harvey

J. Elisha Mitchell Sci. Soc., **42** : 243, 1927.

(Plate 18; Figs. 1-21)

Mycelium dense; principal hyphae stout, straight, less branched, 4-28 μm thick at base, mostly 10-12.5 μm; zoosporangia abundant, terminal, cylindrical, clavate, naviculate, filamentous, sometimes in cluster, 70-210 μm long by 9-30 μm in diameter, mostly 180 × 18 μm; secondary zoosporangia lateral, sympodial; zoospores arranged in 1-5 rows, mostly in 2-3 rows; primary zoospores encyst within the zoosporangium, 5-10 μm in diameter, secondary zoospores liberate by dissolution of sporangial wall, rarely aplanoid; gemmae absent; oogonia abundant, smooth, spherical, 18-40 μm in diameter, mostly 21-25 μm, terminal on lateral branch; antheridial branches diclinous, rarely monoclinous, often lacking on some oogonia; antheridial cells tubular or clavate; oospore single, spherical, eccentric, 12-26.5 μm in diameter, mostly 18 μm (Khulbe, 1985).

Isolated on hemp seed from cultivated soils of Pithoragarh on July 8, 1980, leg. R.D. Khulbe. The type culture has been deposited in the Department of Botany, Kumaun University, Nainital (8/1).

This species was isolated from soil samples (pH 8.6, moisture 26.5%, organic contents 12.6%) of Jeolikot, Nainital on 3rd May 1978 for the first time (Khulbe and Sati, 1979). Later, it was collected from non rhizosphere soils of tomato, paddy, pea and wheat fields near Nainital (Khulbe, 1983b); wheat fields of Nanakmatta, Tarai (Verma, 1984), soils of Ram Tal and Sat Tal lakes (Mer and Khulbe, 1984); cultivated fields of Pithoragarh, Haldwani, Ramgarh and oak forest soils of Naini Tal (Khulbe, 1985). Gupta and Mehrotra (1989a&b) isolated this species from Sanhit pond of Kurukshetra at 14.2-18°C temperature and 7.8-8.1 pH. Recently, this species showed its occurrence in the oak forest soils of Naina Peak (Nainital) during summer and rainy seasons; and in sal forest soils in all the seasons during 14-30°C temperature, 10-90% moisture, 5.3-6.8 pH and 0.8-4.5 organic carbon (Khulbe, 1991).

Brevilegnia indica Prabhuji and Sinha

Neo Botanica 1993, **1** (1&2): 31-38.

(Plate 19; Figs. 1-7)

Mycelium dense, principal hyphae slender, branched, 10-20 μm in diameter at base; zoosporangia abundant, clavate, naviculate or fusiform, 90-270 μm long by 20-50 μm in diameter, renewed sympodially, spore discharge brevilegnoid, encysted spores 10-18 μm in diameter; gemmae absent; oogonia abundant, terminal on main hyphae, on lateral branches of intercalary; sometimes forming glomeruli, oogonial stalks stout or bent or one coiled; oogonia spherical or angular; wall smooth or rarely in 2-4% oogonia papillate, 18-30 μm in diameter; antheridial branches monoclinous, diclinous as well as androgynous; antheridial cells laterally appressed to the oogonium; oospores eccentric, one or very rarely two in an oogonium, 18-23 μm in diameter (Prabhuji and Sinha, 1993).

It was collected from the soil samples of environs of a water body near Chhawari area at Padrauna (District Deoria). U.P., January, 1991.; leg. S.K. Prabhuji. The habitat was characterized by alkaline soil (pH 8.0-8.5), moderate water content (25-30%) and low organic content (3.5-4.5%); type culture (WMSS 1405) has been deposited in the Department of Botany, St. Andrews College, Gorakhpur.

Brevilegnia irregularis Valderrama

J. Elisha Mitchell Sci. Soc. **72**(1), 1956

(Plate 17; Figs. 5-7)

Principal hyphae slender, 15.5-28 µm in diameter; zoosporangia abundant, clavate, fusiform, renewed sympodially, predominantly 160-210 µm by 16-35 µm, generally with 1-4 rows of zoospores, zoospores discharge brevilegnoid, occasionally dictyoid; gemmae absent; oogonia abundant, terminal or lateral, spherical, occasionally pyriform, subglobose or irregulare; 18-36 µm in diameter, sparsely papillate, inner surface of wall usually irregular; oospore generally maturing, eccentric, single, spherical, not filling the oogonium, 12.5-32 µm in diameter; antheridial branches usually monoclinous, very rarely androgynous.

This species was collected by Khare (1992) from Jalalpur lake, Bareilly.

Brevilegnia linearis Coker and Braxton

Jr. Elisha, Mitchell., Sci. Soc. **42** : 214-215, 1927

(Plate 20; Figs. 1-10)

Mycelium limited, dense, opaque; principal hyphae slender, 7-19 µm thick at base, mostly 12.5-15 µm; zoosporangia abundant, terminal on the main hyphae or lateral branches, occasionally intercalary, long, slender, simple, rarely branched, 180-700 µm long by 8-17 µm in diameter, mostly 250-335 µm × 12-15 µm, renewed sympodially; zoospores arranged in a single row, 2-40 in number, mostly 15-20, discharged by disintegration of sporangial wall, spherical to elongate, 7-14 µm in diameter, 10-18 µm in length, swimming stage absent; gemmae absent; oogonia abundant, lateral, spherical, 18-38 µm in diameter, mostly 25-35 µm; oogonial wall smooth, thin, very rarely pitted under point of attachment of antheridial cell; oogonial stalk often bent variously; antheridial cell short, tuberous, generally androgynous, rarely diclinous or monoclinous often absent on some oogonia; oospore single, spherical, plerotic, occasionally aplerotic, eccentric, 15-35 µm in diameter, mostly 22-31 µm.

Isolated on hemp seed from oak forest of Nainital on April 17, 1980, leg. R.D. Khulbe. The type culture has been deposited in the Department of Botany, Kumaun University, Nainital (8/2).

This species was isolated from Nainital lake on 6 March 1973 for the first time (Khulbe, 1977 & 1980a) and found to be distributed in the water samples of Sariyatal, Khurpatal, Bhimtal and Naukuchiya Tal lakes during winter and spring when temperature ranged from 10-20° C (Khulbe and Bhargava, 1977) and pH 7.4-7.9 (Khulbe, 1980b).

This species was again collected from soil samples in the vicinity of Ram Tal and Nal Damyanti Tal (Naini Tal) during autumn season at 16-20°C temperature (Mer *et al.*, 1980); forest soils of Kilbury, Ramgarh and at Naina peak (Nainital) during spring and winter seasons only at 5°C temperature, 5.9-6.6 pH, 32-85% moisture, 1.8 -3.0% organic carbon (Khulbe, 1985&1991).

Brevilegnia megasperma Harvey

J. Elisha Mitchell Sci. Soc. **45** : 322, 1930

(Plate 21; Figs. 1-12)

Mycelium dense, white, reaching a ring growth of 10-15 μm within 3-4 weeks on hemp seed; hyphae thick, irregular, sparingly branched, 30-65 μm thick at base; zoosporangia terminal on main hyphae, clavate, filamentous, 15-38 μm thick, 20-300 μm long; secondary sporangia formed at the tip of hyphal branches which arise by lateral proliferation; spores broad, spherical, slightly angular, formed in single or many rows, encyst in situ and escape after decay of sporangial wall, 8-12 μm thick; gemmae abundant in some cultures, dark, dense, size variable, may give rise to sporangia or sprout into hyphal filaments; oogonia abundant, lateral on short to long stalks, spherical, occasionally ovate, smooth, sometimes irregular, 22-60 μm in diameter; oogonial wall about 1.5 μm thick; oospore single, spherical, 15-50 μm in diameter, eccentric; antheridia usually absent, but sometimes present in hypogynous condition (Khulbe, 1985).

Isolated on hemp seed from paddy field soils of Pithoragarh on June 20, 1983, leg. R.D. Khulbe. The type culture has been deposited in the Department of Botany, Kumaun University, Nainital (B/3). This species was also isolated from cultivated soils of Lohaghat.

Plate-17

Figs. 1-4. *Brevilegnia bispora* : 1-2; Zoosporangia showing internal encystment of zoospores.

3-Oogonium with papilla and mature single, eccentric oospores.

4-Lateral oogonium with monoclinous antheridium. (Misra, 1980).

Figs. 5-7. *B. irregularis* : 5 & 6-Oogonia with single eccentric oospores, monoclinous and diclinous antheridia.

7-Zoosporangium showing single row of zoospores. (Khare, 1992).

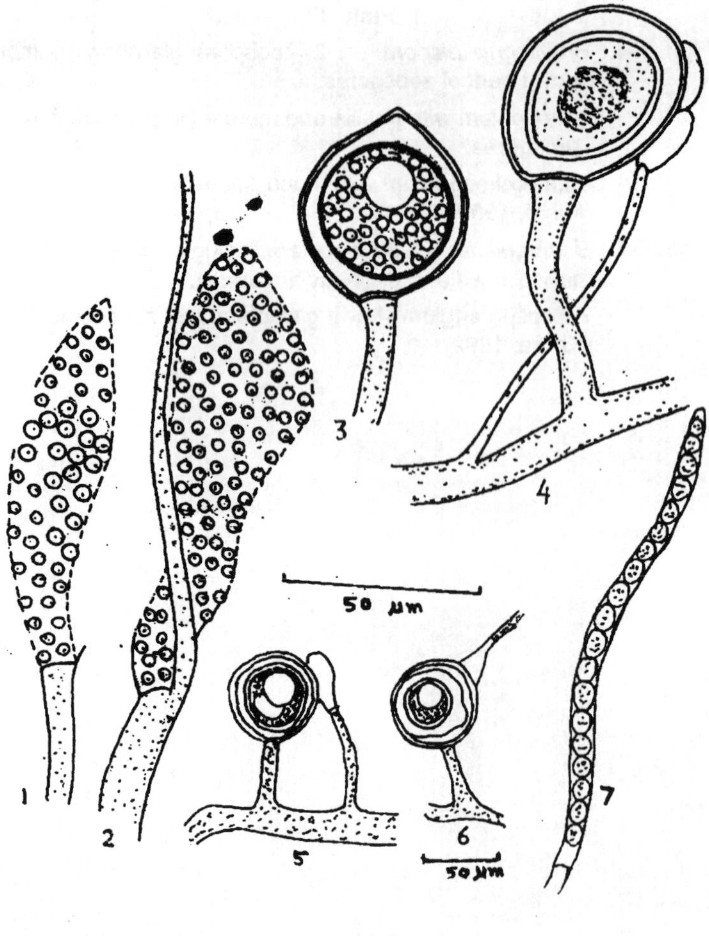

Plate 17

Plate-18

Figs. 1-21 : *Brevilegnia diclina* Harvey.

1. Single hyphal tip
2. Terminal young zoosporangium.
3. Mature zoosporangium showing liberation of zoospores.
4. Long, cylindrical zoosporangium.

5-10. Different types of zoosporangia.

11. Small zoosporangium with 2 zoospores.
12. Zoosporangium with variable zoospores.
13. Curved zoosporangium.
14. Hypha having two types of zoosporangia—one is cylindrical with 2 rows of zoospores and the other is filamentous, having single row of zoospores.
15. Zoosporangium having one and two rows of zoospores.
16. Long simple filamentous zoosporangium having single row of zoospores.
17. Branched zoosporangium.

18-20. Young oogonia with diclinous antheridial branches.

21. Mature oogonia with single eccentric oospores and diclinous antheridial branches.

Figs. 1-4, 14-17 & 20, scale A;

Figs. 5-13, 18 & 19, scale C;

Fig. 21, scale B.

(Khulbe, 1985)

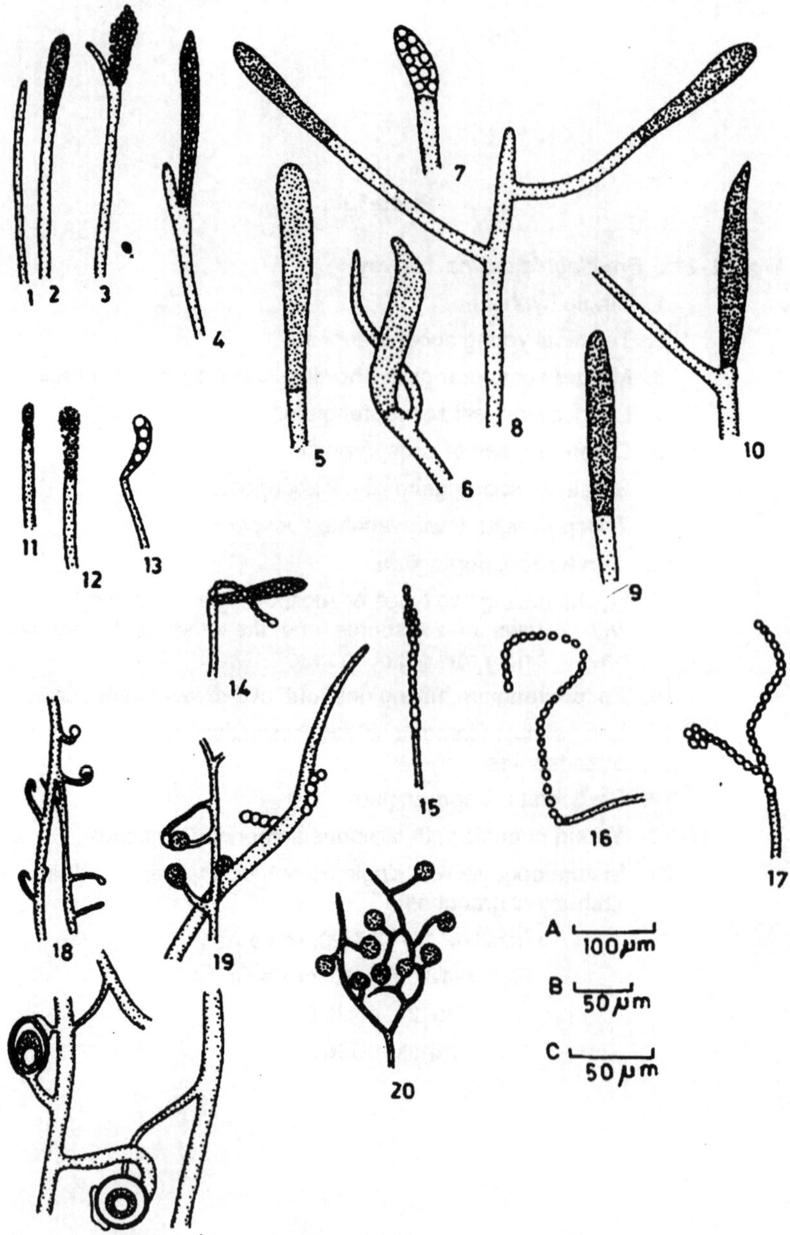

Plate 18

Plate-19

Figs. 1-7. *Brevilegnia indica* : 1-Large sized sporangium showing typical brevilegnoid spore discharge.

2- Mature terminal sporangium before dehiscence.

3- Oogonium with single eccentric oospore showing long oogonial stalk, diclinous and androgynous antheridia.

4- Glomerule of oogonia with monoclinous and androgynous antheridial branches.

5- Terminal oogonium having papillate oogonial wall with androgynous antheridium.

6. Intercalary oogonium showing papillate oogonial wall with diclinous antheridia.

7- Oogonium with two oospheres

(Prabhuji & Sinha, 1993)

Figs. 1-2 scale b, Figs. 3-7 scale a

(Prabhuji & Sinha, 1993).

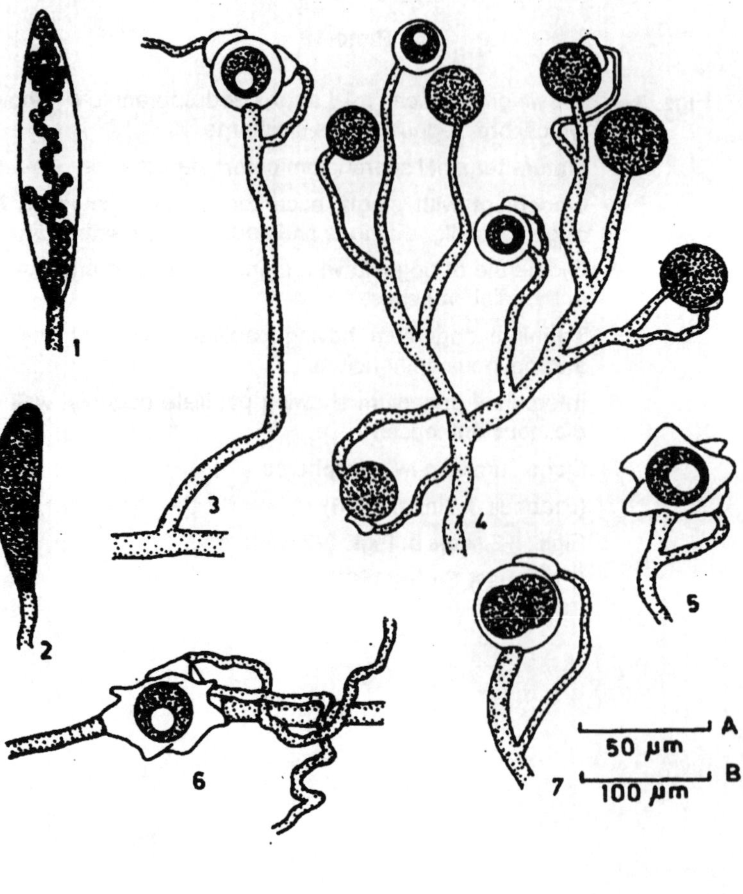

Plate 19

Plate-20

Figs. 1-10 : *Brevilegnia linearis* Coker & Braxton.

1. Hypha showing terminal zoosporangium with a row of zoospores.
2. Lateral zoosporangium
3. Terminal zoosporangium and lateral stalked oogonia.
4. Zoosporangium with cylindrical and spherical zoospores.

5-7. Hyphae having lateral stalked oogonia.

8. Oogonium with androgynous antheridium.
9. Coiled oogonial stalk.
10. Oogonium with diclinous antheridium.

Figs. 1-10, scale A.

Figs. 11-30 : *Brevilegnia subclavata* Couch.

11-14. Different types of zoosporangia.

15&16. Zoosporangia with mature zoospores in many rows.

17. Terminal and basipetal young zoosporangia.

18-21. Different types of zoosporangia showing single row of zoospores.

22. Hypha showing terminal two subspherical zoosporangia with many zoospores.
23. Intercalary zoosporangium with similar zoospores.
24. Intercalary zoosporangium with different types of zoospores.
25. Encysted zoospores.
26. A hypha showing terminal mature zoosporangium and lateral young oogonia.
27. Terminal oogonium with single, eccentric oospores without antheridium, and a lateral oogonium with androgynous antheridium.
28. Lateral oogonium without antheridium.
29. Lateral oogonium with diclinous antheridium.
30. Basipetal succession of zoosporangia.

Figs. 11-14, 26 & 30, scale A; Figs. 25, 27-29, scale B.

(Khulbe, 1985)

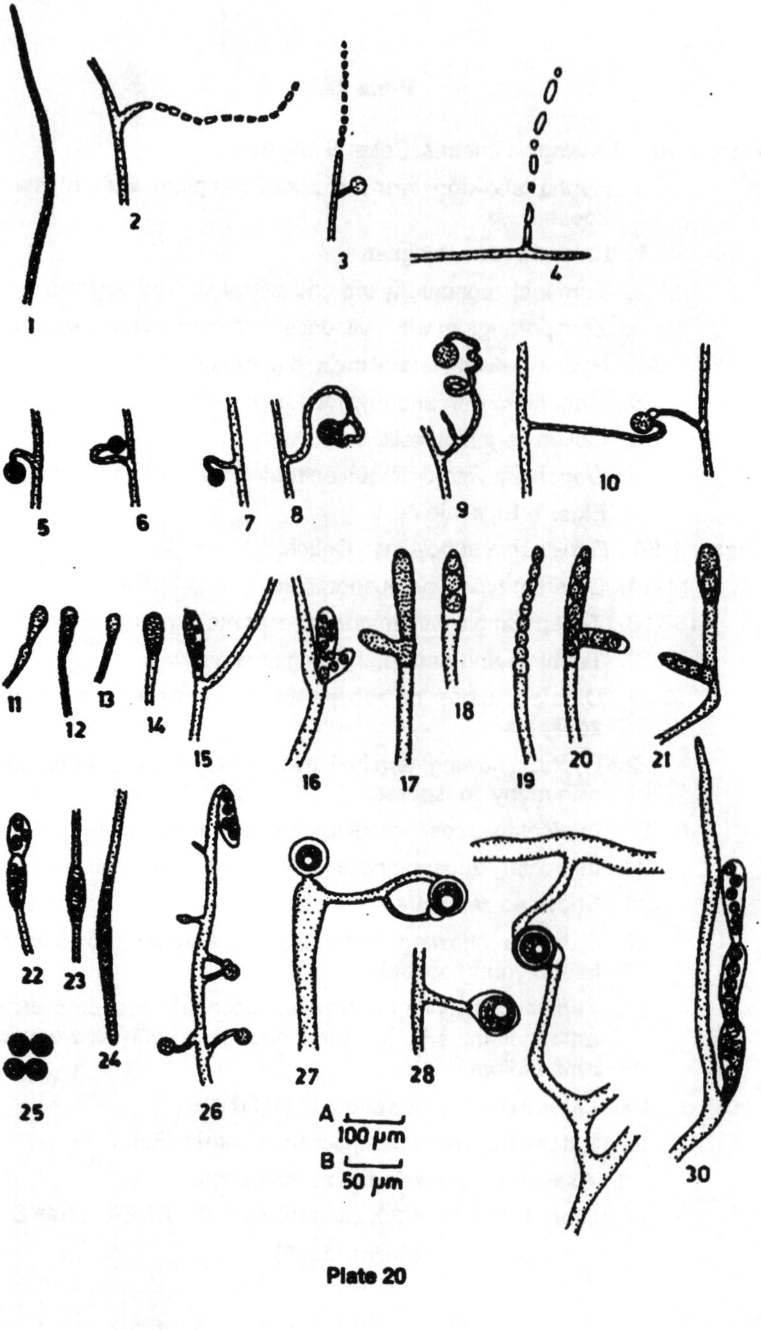
1
2
3
4
5
6
7
8
9
10
11
12
13
14
15
16
17
18
19
20
21
22
23
24
25
26
27
28
30
A
100 µm
B
50 µm

Plate 20

Plate-21

Figs. 1-12 : *Brevilegnia megasperma* Harvey.

1-2. Hyphal tips.

3 & 4. Terminal young zoosporangia.

5 & 6. Showing zoospores liberation and secondary zoosporangia.

7. Intercalary zoosporangium.

8. Gemmae.

9. Encysted zoospores.

10. Hypha bearing two young oogonia and one mature oogonium with eccentric oospore.

11. An oogonium with hypogynous antheridium.

12. Mature oogonium with single eccentric oospore.

All Figs., same scale.

(Khulbe, 1985)

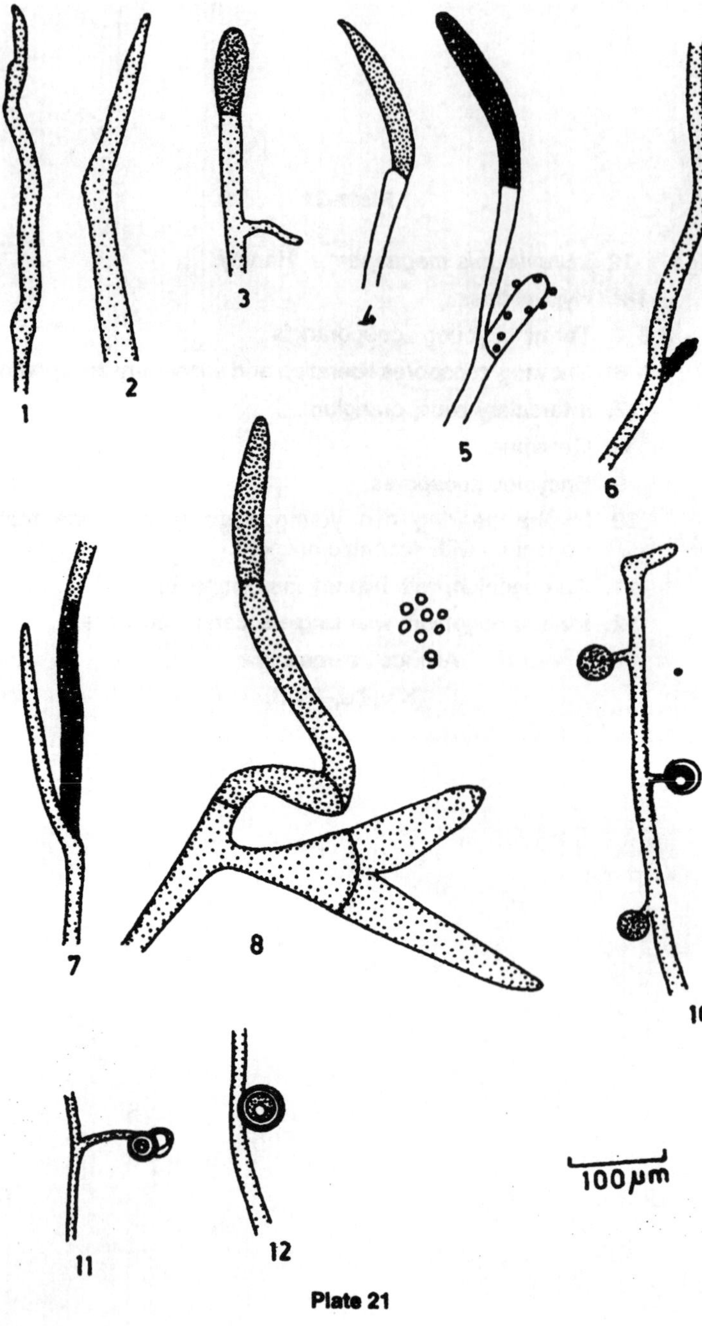

Plate 21

Brevilegnia subclavata Couch

J. Elisha Mitchell Sci. Soc., **42** : 229, 1927

(Plate 20; Figs. 11-30)

Mycelium dense, principal hyphae short, straight, sparingly branched, 15-65 μm thick at base, mostly 25-35 μm; zoosporangia abundant, terminal, cylindrical, clavate, 50-180 μm long by 15-80 μm in diameter, mostly 80-90 x 30-50 μm, renewed sympodially as well as in basipetal succession of 2-9; zoospores in 1-5 rows, 8-24 μm in diameter, spherical to cylindrical, swimming stage absent, gemmae absent; oogonia abundant, lateral, spherical, subspherical, 18-48 μm in diameter, mostly 25-35 μm; antheridia monoclinous, rarely diclinous and often lacking; oospore single, eccentric, 14-36 μm in diameter, generally 18-25 μm (Khulbe, 1985).

Isolated on hemp seed from forest soils of Haldwani on March 15, 1980, leg. R.D. Khulbe. The type culture has been deposited in the Department of Botany, Kumaun University, Nainital (B/4). The present species was also found to be distributed in the forest soils of Kausani and Ramgarh.

The species was isolated from silty soils of Jalwania pond, Gorakhpur for the first time (Prabhuji and Srivastava, 1978a, Prabhuji, 1979 & 1984). They observed best growth of the species on Dextrose-Peptone Agar at 25-28°C, but not on snake skin and other keratinophilous substrates. Later, this species was collected from cultivated soils of Khatima (Verma, 1984), pine forest of Kausani, oak forest of Ramgarh; and in sal forest soils of Haldwani in all the seasons except winter at 14-30°C, 5.8-6.8pH, 2-26 per cent moisture and 0.4-4.2 per cent organic carbon (Khulbe, 1985 & 1991).

Dictyuchus Leitgeb

Bot. Zeit. **26** : 502, 1868

The Saprolegniaceae, 150, 1923

Thalli fertile or sterile; fertile ones dioecious, hyphae slender or stout, simple or branched; tips of hyphae blunt; 8-75 μm thick at base; primary zoosporangia terminal, cylindrical, blunt, secondary ones are lateral, rarely intercalary, deciduous, primary zoospores encyst within the zoosporangium;

100-1000 × 9-50 μm in size; secondary zoospores release from separate pores formed in the walls, leaving primary walls of the zoospores within zoosporangium and forming their a network of walls; secondary zoospores undergo encystment and germinate by germ tubes, some of them again emerge and swim before sprouting and show polyplanetism; encysted zoospores 8-17 μm in diameter; gemmae rare, lateral or intercalary, behave as zoosporangium or single spore; antheridia diclinous or androgynous or absent; oogonia lateral, spherical, smooth, unpitted or absent; 27-43.8 μm in diameter; oospore one, aplerotic, eccentric, 18-41 μm in diameter.

Eight species of *Dictyuchus* have been reported from India. Their distinguishing characters are given in the following key:

Sex organs absent .... *D. sterile*, p. 135

Sex organs present .... A

A_1 Antheridia present.. B

A2 Antheridia absent C

B_1 Antheridia rare, diclinous *D. lucknowensis*, p. 132

B_2 Antheridia abundant.. D

C_1 Oogonia 28-43.8 μm, oospores 18-32 μm, aplerotic.. *D. anomalus*, p. 131

C_2 Oogonia 27-34 μm, oospore 20-27 μm .. *D.missouriensis*, p. 132

D_1 Antheridia strictly diclinous .... E

D_2 Antheridia monoclinous, diclinous or androgynous .... *D. pseudodictyon*, p. 135

E_1 Antheridia often enwrape the oogonia... F

E_2 Antheridia not enwrape the oogonia.... *D pisci*, p. 134

F_1 Oospore single, centric.. *D. monosporus*, p. 133

F_2 Oospore single, eccentric .... *D. carpophorus*, p. 131

Dictyuchus anomalus Nagai

J. Fac. Agr. Hokkaida, Imp. Univ. *32* : 1-43, 1931

(Plate 22; Figs. 11-13)

Hyphae stout, 21.0-37.5 μm in diam., mostly unbranched, rarely branched near the tip, branches short or long bearing sporangia at the tip; sporangia numerous, cylindrical, 120-544.8 μm long by 22.5-38 μm thick, mostly 300-375 × 28-33 μm, terminal, rarely intercalary; spores arranged inside the sporangia in a typical manner, often germinating within the sporangium by producing a long germ tube (at 25-30°C); oogonia terminally and singly borne on the branches of oogonial stalks (47.19-109.30 × 5.9-7.24 μm), straight or curved, 2-3 branched per oogonial stalk; oogonia smooth, globular, unpitted, 28-43.8 μm in diam. with a single oospore; oospore smooth, globular, eccentric, 18.0-32.67 μm in diam., mostly 25-41 μm in diam., aplerotic, antheridia not observed (Srivastava, 1981).

This species was isolated from a dieased fish of Ramgarh lake Gorakhpur (Srivastava, 1981); lake of Hyderabad (Manoharachary, 1981) and from paddy field of Nanakmatta on July 10, 1981 (Verma, 1984); Jalalpur lake, Bareilly club pond, Badaun road pond and Dhopeshwar pond, Bareilly (Khare 1992).

Dictyuchus carpophorus Zopf

Beitrage Z. Phys U. Morph. n. Organisms **3:** *48* : 1893

(Plate 22; Figs. 1-4)

Growth thick, two week old colony approximately 1.5 cm in diam. with main hyphae 16-85 μm thick at base; sporangia terminal, lobed, long, clavate or slightly curved, 169-592 μm long by 19-37 μm broad, often containing a single row of spores; gemmae absent; oogonia spherical or pyriform, abundant terminal, single on long or short branches of the main threads, 32-56 μm in diameter, the wall pitted only under the point of antheridial attachment, otherwise smooth and rarely proliferating; antheridia always present, usually several to each oogonium, borne on slender branches of diclinous origin which often completely enwrape the oogonium; oospore single, spherical, smooth, eccentric, 18-41 μm in diam., predominantly 30 μm;

germination of oospores not observed (Dayal and Thakurji, 1968b).

Isolated on hemp seed from pond near agricultural farm, Varanasi, leg, Thakurji, 6 Nov. 1966, slide RD 15 (Dayal and Thakurji 1968b) and from alkaline water and mud around Lucknow (Misra, 1982). Khare (1992) isolated it from Nakatiya river, Bareilly club pond and Dhopeshwar pond, Bareilly.

Dictyuchus lucknowensis Rai and Misra

Current Science **47** (5): 171-172, 1978

(Plate 22; Figs. 5,6,14)

Vegetative growth moderately dense on hemp seeds; hyphae branched; gemmae present in old cultures, formed by the regeneration of the hyphae; sporangia achlyoid as well as dictyoid, former rare, latter ones abundant, and are of true net type, dictyoid type of sporangia branched, all arising from the tips of hyphae and by cymose branching, 167-655 × 11-33 µm; spores 8-17 µm in diameter; oogonia abundant, spherical, borne singly on lateral branches and measuring 25-42 µm in diameter; oogonial wall unpitted and smooth; antheridia rarely present, one per oogonium if present and diclinous; oospore one, 22-28 µm in diam., eccentric.

Isolated from a pond (pH 8.5) Baba Ka Pokhara, Lucknow (Misra, 1982).

Dictyuchus missouriensis Couch

J. Elisha Mitchell Sci. Soc. **46** : 227, 1931

Proc. Nat. Acad. Sci. India **38**(1) : 32-38, 1968

(Plate 22; Figs. 8-10)

Growth thick, broad with hyphae up to 24 µm thick, tips of hyphae rather blunt; primary sporangia nearly cylindrical, blunt, borne typically in a zig-zag sympodium with long internodes; later, they are formed by the segmentation of the hyphae into the long joint and in such case after the spores are formed, rest like gemmae and exhibit a strong tendency to fall away from each other and from the hyphae and to lie free in the water; sporangia terminal, long, little broad with 17-34 µm in thickness containing more than single row of spores, proliferating by seperated sprouting from the side below to form sympodium; spores not escaping from the sporangium and forming there net work of walls, from which they emerge after a rest by individual opening to the outside; gemmae not represented unless resting sporangia with spores in them be considered such; oogonia spherical, oval, smooth; wall unpitted; terminal on long slender branches up to 27-34 µm in diam., predominantly about 29 µm; antheridia not formed; oospore single, smooth, spherical, eccentric, filling almost entire inner space of an oogonium, 20-27 µm in diam., mostly 21 µm, oil globule one, large 10-17 µm in diam.; germination not observed (Dayal and Thakurji, 1968 b).

Isolated on hemp seed from pond water near Manduadih, Varanasi, leg., Thakurji, 20 Feb. 1964, Slide RD 60 (Dayal and Thakurji, 1968a) and from water and soil around Lucknow (Misra, 1980), and Sagar, M.P. (Saksena and Rajgopalan, 1958).

Dictyuchus monosporus Leitgeb

N. Amer. Flora, **2**(1): 1-76, 1937.

Lloydia **4** : 270-273, 1941.

(Plate 22, Figs. 15-16)

Principal hyphae moderately stout, branched, tapering, 8-75 µm thick at base, mostly 25-45 µm; zoosporangia abundant, subcylindrical, cylindrical, renewed sympodially, rarely basipetal or intercalary, 135-1000 µm long by 9-50 µm in diameter, mostly deciduous; zoospores in 2-5 rows, encysted within the zoosporangium, secondary zoospores liberate separately from individual pores leaving the wall of primary zoospore in the zoosporangium and give appearance of true net, encysted zoospores 9-10.5 µm in diameter; gemmae and sex organs absent (Khulbe, 1985).

Isolated on hemp seed from forest soil of Haldwani on October 15, 1980, leg. R.D. Khulbe. The type culture has been deposited in the Department of Botany, Kumaun University, Nainital (D/1).

This species was isolated from dead plant fragments under water from Dehradun (Butler, 1907); water of Gorakhpur (Srivastava, 1967a, Prabhuji, 1979); sandy, silt and clayey soils of the environs of Laldiggi pond, Ramgarh lake and Asuran-ka-pokhra pond of Gorakhpur (Prabhuji and Srivastava, 1977); from Pune (Kapadnis, 1980); soils around Nainital (Sati and Khulbe, 1980); as a parasite of some fishes (Srivastava and Mekrani, 1980a, Sati, 1981); soil and water samples of IARI, New Delhi, during winter at 10-25°C temperature (Chowdhry and Aggarwal, 1980 a&b); alkaline waters of Lucknow (Misra, 1982) ; soil roots, and seedlings of rice in Khatima, Nainital (Khulbe and Verma, 1983; Verma, 1984); Brahma and Sanhit Sarovar of Kurukshetra (Gupta and Mehrotra, 1988a, 1989 a&b).

In the recent past, this species was collected from Sal forest soils of Haldwani during autumn season only at 25°C temperature, 12-18 per cent moisture, 6.1-6.2 pH and 1.8-2 per cent organic carbon (Khulbe, 1991). Cultivated soils of Dwarahat, Ranikhet and Pithoragarh and water and soil samples of Sat Tal lake also exhibited this species (Khulbe, 1991).

Dictyuchus pisci Khulbe and Sati

Bibliotheca Mycologica **91** : 609-617, 1983

(Plate 23, Figs. 1-6)

Mycelium fairly vigorous on hemp seed; diameter of 14 days old colony 2-3 cm; hyphae up to 32 µm thick at the base; gemmae lacking; zoosporangia abundant, simple, rarely branched, persistant, rarely deciduous; primary zoosporangia on the tip of the main hyphae, long, cylindrical to clavate, 19-31 µm × 225-375 µm in size, mostly 19 × 280 µm; zoosporangia sympodial, rarely cymose; zoospores liberate through lateral papillae leaving true-net type zoosporangia; encysted zoospores 13-21 µm in diameter; oogonia abundant, sometimes abortive, borne on slender stalk of variable length, usually long, lateral on the hyphae, spherical or subspherical, 25-50 µm in diameter, mostly 25 µm; oogonial wall smooth, unpitted except under antheridial attachment; antheridia diclinous, 1-3 per oogonium, sometimes completely enwrapping the oogonium; oospore single, spherical, eccentric, 23-45 µm in diameter, mostly 23 µm (Khulbe, 1985).

This species was isolated from infected *Carassius auratus* of a fish pond at Jeolikot, Nainital on 14-2-81, and deposited at CMI, Kew, England (IMI-264216).

Dictyuchus pseudodictyon Coker and Braxton

J. Elisha Mitchell Sci. Soc. **42** : 139-149, 1926.

(Plate 22; Fig. 7)

Mycelium limited, diffuse, moderately branched; gemmae abundant, filiform; sporangia abundant, fusiform, measuring 100-400 × 15-30 µm, sympodially branched; zoospores discharge dictyoid, true net type, very occasional false net type, achlyoid type of discharge not seen; oogonia abundant, lateral, spherical, 30-33 µm in diameter, smooth, pitted; oogonial stalk curved, bent; antheridial branches monoclinous or diclinous, giving umbratiform appearance; antheridial cells presistent; oospore one per oogonium, spherical, 19-28 µm in diameter, eccentric, mature oospore germinating by short unbranched germ tube, bearing small filiform true net type of dictyoid sporangia.

This species was isolated from pond water (pH 8.1), broad tank of fisheries farm, Utrethiya, Lucknow (Misra, 1980).

Dictyuchus sterile Coker

The Saprolegniaceae p. 151, pl. 52, 1923

(Plate 24; Figs. 1-7)

Hyphae stout, much branched, 8-75 µm in diameter, predominantly 26-45 µm, primary sporangia formed at the tips of the hyphae, rarely intercalary, cylindrical, blunt, renewed sympodially, branching in sporangia is also of common occurrence, 195-840 µm long and 9-41 µm in diameter, predominantly 30-33.8 µm, mostly deciduous; primary zoospores arranged in one to three rows, not liberating from the sporangium, but remaining within the zoosporangium and forming a net work of walls separating each zoospores as a separate entity in a separate cell, liberate by individual opening, sometimes germinate within the zoosporangium as in aplanes; encysted zoospores 11-15 µm in diameter, gemmae absent; sex organs not developed.

This species was isolated by Mer (1982) from Ram Tal, Sat Tal and Nainital on January 1, 1979. Mitra (1935), Srivastava (1967), Prabhuji (1979) and Srivastava and Mekrani (1980) isolated it from warmer habitats of India.

Plate–22

Figs. 1-4. *Dictyuchus carpophorus* :1-A hypha with young and mature sporangium showing dictyoid liberation of zoospores. 2-branched typical sporangium showing true net. 3-oogonium with antheridium and single oospores. 4-oogonium with long stalk and diclinous antheridium, eccentric single spherical oospore (Figs. 2&4-Dayal & Thakurji, 1968b).

Figs. 5,6 and 14. *D. lucknowensis.* 5-sporangium showing achlyoid liberation of zoospores.

6-mature zoosporangium with liberation of zoospores and true net, 14-lateral oogonium with monoclinous antheridium and single oospore (Misra, 1980).

Fig. 7. *D. pseudodictyon* -lateral oogonium with monoclinous antheridia, single oospore (Misra, 1980).

Figs 8-10. *D. missouriensis* -8-hypha showing many lateral young zoosporangia, 9-oogonium with long stalk, 10-oogonium with small curved stalk (Dayal & Thakurji, 1968b).

Figs. 11-13 *D. anomalus.* 11-hyphae, 12 & 13-hyphae with lateral oogonia and no antheridium (Srivastava, 1981).

Figs. 15&16 *D. monosporus.* hyphae with zoosporangia showing dictyoid spore liberation (Misra, 1980).

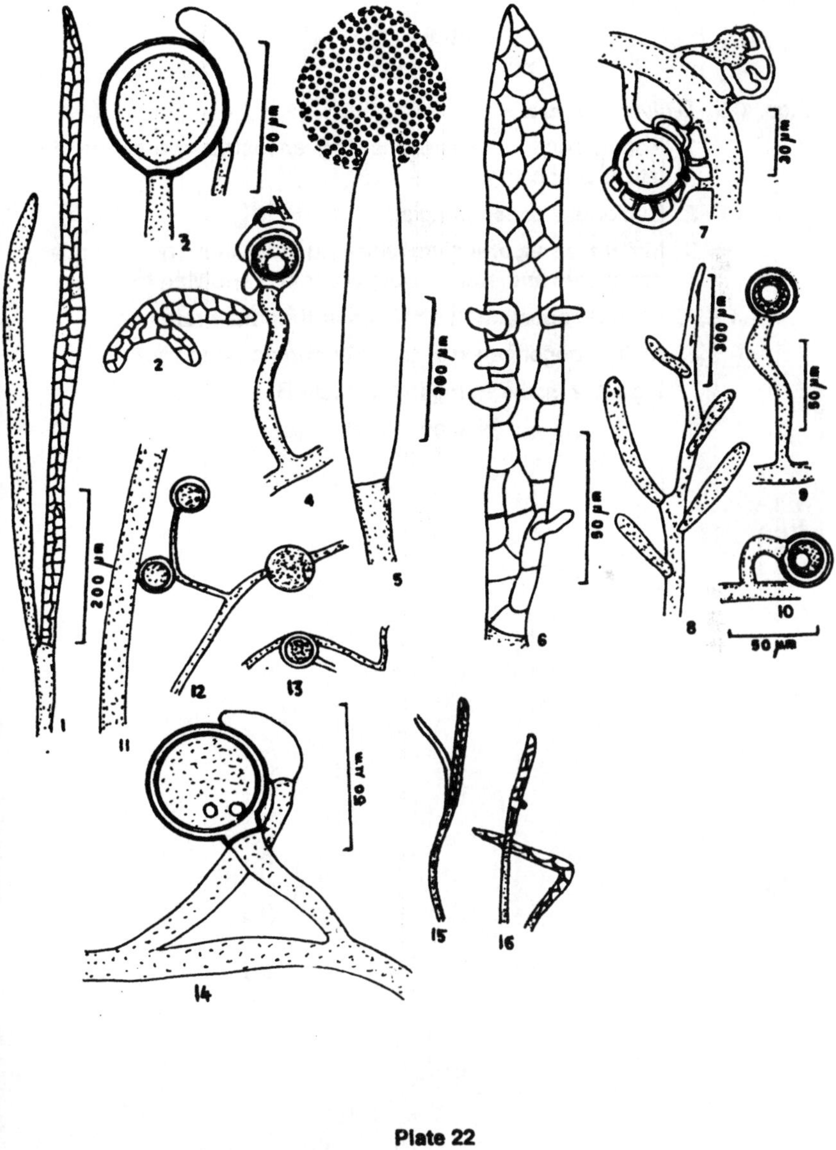
1
2
3
4
5
6
7
8
9
10
11
12
13
14
15
16
50 μm
300 μm
50 μm
30 μm
300 μm
50 μm
50 μm
200 μm
50 μm

Plate 22

Plate–23

Figs. 1-6: *Dictyuchus pisci* .

1. Young zoosporangium showing encysted zoospores and lateral branching.
2. Deciduous zoosporangia.
3. Mature zoosporangium showing liberation of secondary zoospores and empty zoosporangium showing true-net.

4 & 5. Lateral oogonia with diclinous antheridial branches.

6. Mature oogonium with single eccentric oospore.

Figs. 1-2, scale A; Figs. 3-6, scale B.

(Khulbe, 1985)

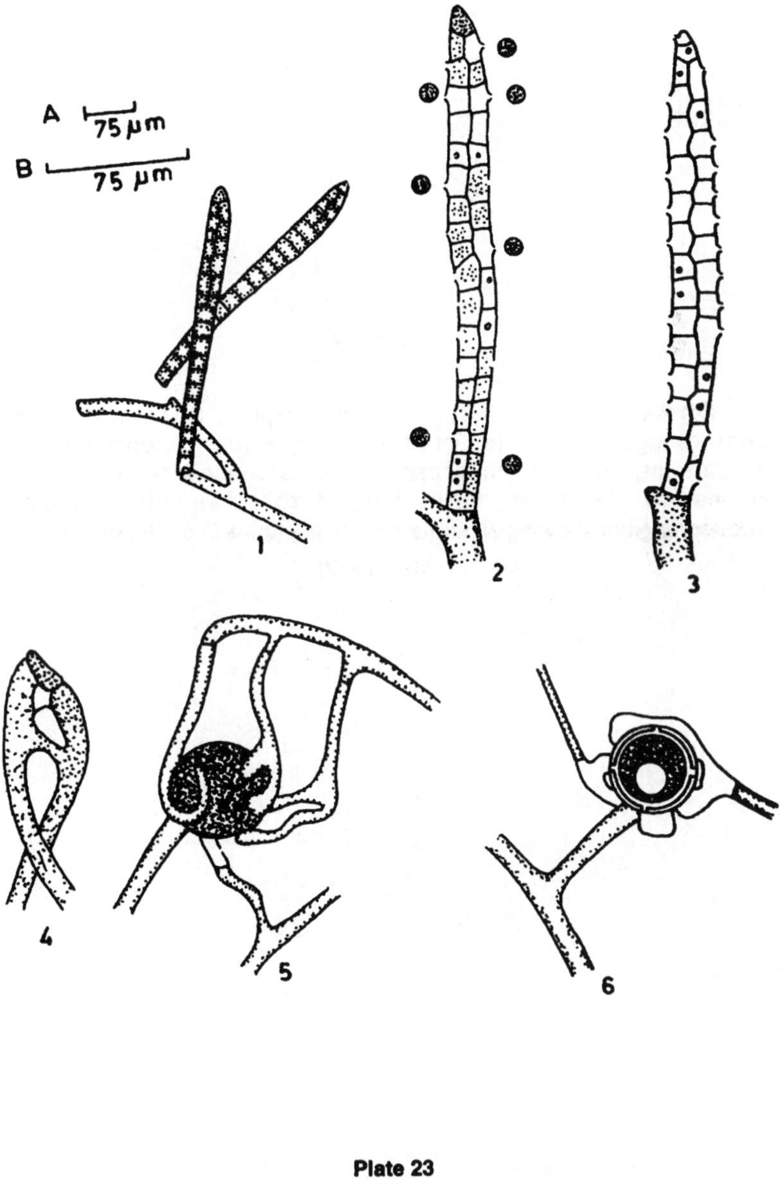

Plate 23

Plate–24

Dictyuchus sterile : 1, Young sporangium; 2. showing mature zoosporangia; 3, mature, intercalary single rowed zoosporangium; 4, showing liberation of zoospores; 5, branched zoosporangium; 6, mature, deciduous, sickle shaped zoosporangium; 7, empty zoosporangium showing separate individual papilla in each compartment.

(Mer, 1982)

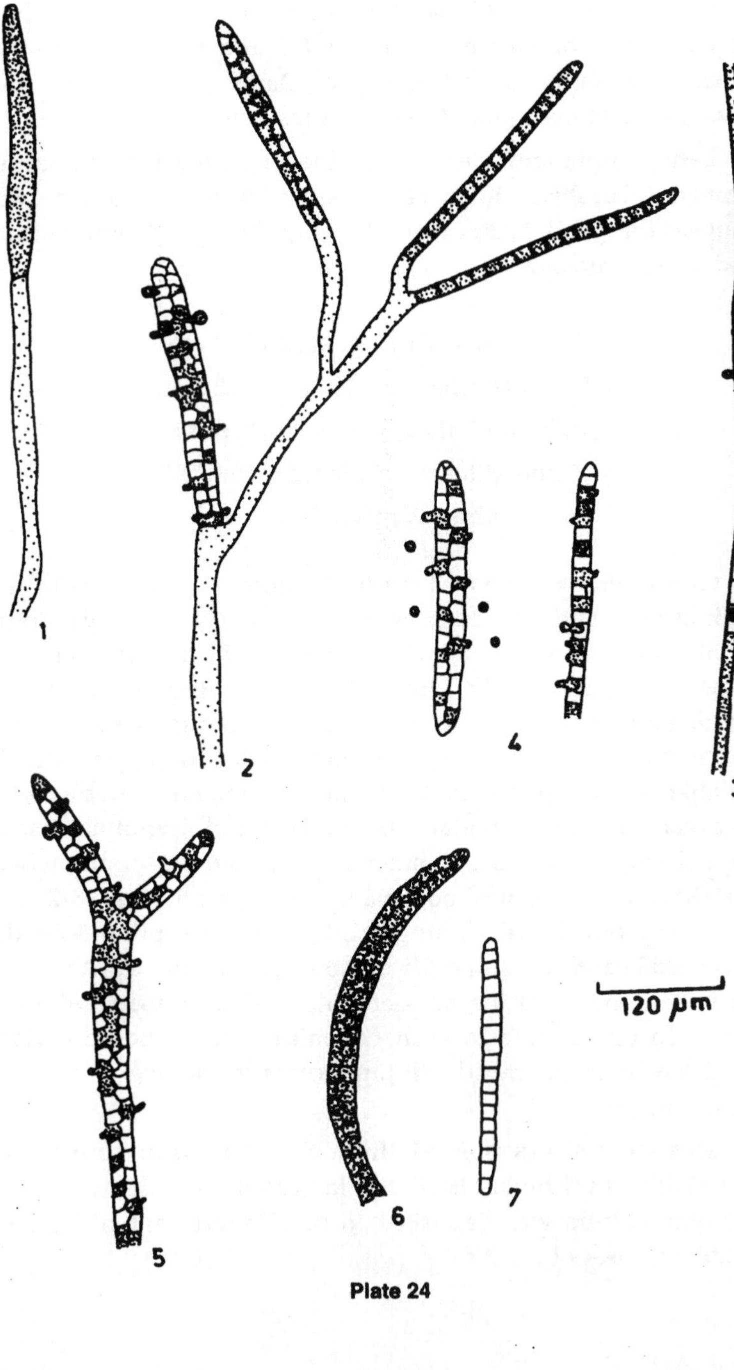

Plate 24

Khulbe and Sati (1983) isolated *Dictyuchus sterile* from infected *Tor tor, Barbus chillinoides, Puntius ticto* and *Carassius auratus* fishes from Bhim Tal, Nainital, Nal Damyanti Tal lakes and Government House pond, Nainital, respectively.

Later, Gupta and Mehrotra (1988a) reported it from Sanhit Sarovar, Kurukshetra. Khare (1992) isolated it from Nakatiya river, Dhopeshwar pond, Jalalpur lake, Bareilly Club pond, and Badaun road pond, Bareilly.

Geolegnia Coker and Harvey

J. Elisha Mitchell Sci. Soc. *41* : 153, 1925

Geolegnia inflata Coker and Harvey

J. Elisha Mitchell Sci., Soc. *41*: 154, 1925

(Plate 25; Figs. 1-12)

One week old colony 2 cm in diameter. Hyphae 10-16 µm thick at base and 2.5 µm thick at tip. Primary sporangia long, inflated, unbranched, terminal, constricted, 150-450 µm long and 12-20 µm thick, mostly 225 × 15 µm showing characteristic swelling about 15-22 µm diam. Spores arranged in one row, non-motile, spherical, rarely oval, usually similar in shape, slightly variable in some sporangia, 13-15 µm in diam., released by decay of sporangial wall. Secondary sporangia lateral. Gemmae absent. Oogonia many, lateral on short to long, simple or branched, straight or curved stalks; oogonia spherical, pyriform, 15-20 µm in diam., mostly 16.5 µm; smooth, only one pit below the antheridial contact. Antheridia on all oogonia, androgynous, 1-2 per oogonium, tuberous, swollen, laterally appressed, 5-6.8 µm thick. Oospore single in each oogonium, aplerotic, spherical, 13-18 µm in diam., mostly 16 µm, eccentric, having a large oil globule at one side.

This species was isolated from *Shorea robusta* forest soils near Haldwani (Khulbe, 1982) and indicated its endemic nature. The type culture was deposited in the Department of Botany, Nainital (RDK-G 1).

Plate-25

Figs. 1-12 : *Geolegnia inflata* Coker & Harvey.

1. Single hyphal tip.

2&3. Zoosporangial development.

4&7. Mature zoosporangia showing characteristic swellings.

5,6&8. Liberation of encysted zoospores.

9. Intercalary gemma.

10&11. Young oogonia with monoclinous antheridial branches.

12. Mature oogonium with single eccentric oospore and androgynous antheridium.

All Figs., same scale.

(Khulbe, 1985)

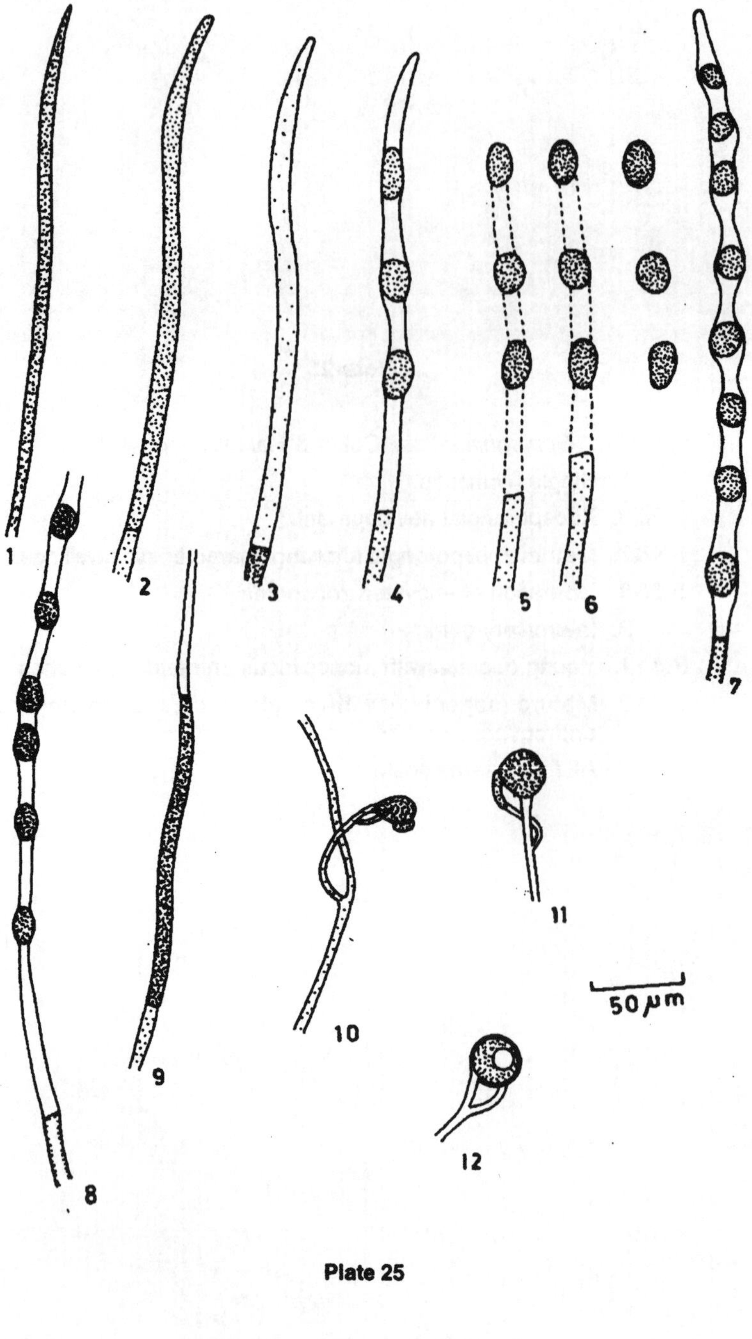

Plate 25

Hamidia Chaudhuri

Proc. Indian Acad. Sci. *15*, 4B : 225, 1942

Hamidia indica **Chaudhuri**

Proc. Indian Acad. Sci. *15*; 4B : 227-230, 1942

(Plate 26; Figs. 1-10)

Hyphae hyaline, generally racemosely branched, delicate, sparsely septate, bearing oogonia and sporangia both in racemose and cymose manner, best seen in natural condition; oogonia borne singly or in clusters (of usually 3); oogonia when borne singly have long stalks which break at the slightest disturbance; oogonium with a single large egg entirely filling it. Egg with smooth wall, emerges gradually and imperceptibly; apandrous; sporangia resembling oogonia also formed, each with 2-7 swarm spores; discharge tube may or may not be present; swarm spores, without a properly differentiated wall and non ciliate (monoplanetic), germinate soon after discharge; undischarged swarm spores germinate inside the sporangium, putting forth unilaterally or bilaterally produced germ tubes; gemmae also formed; growth on fresh potato stalks most luxuriant; egg yolk and insects unsuitable for growth (Chaudhuri *et al*., 1947).

Collected in March 1936 from decaying twigs under water from Tarn Taran and Barhamjit-Punjab. Type specimen was deposited in Punjab University Botany Department, Lahore (Chaudhuri, 1942; Chaudhuri *et al.*, 1947). The species was cultured on potato blocks. This species has not been reported again.

Isoachlya Kauffman

Amer. J. Bot. **8** : 231, 1921

The Saprolegniaceae, 81, 1923

Hyphae slender, branched, 8 to 35 µm thick at base; primary zoosporangia terminal, oval, pyriform, elongated, clavate, cylindrical; secondary ones lateral develop mostly by cymose or pseudo-cymose manner as in *Achlya*, some or rarely by internal proliferation; or completely lateral; size 14.5-45 × 99-500 µm, zoospores diplanetic, escape and swarm as in *Saprolegnia*; encysted

Plate-26

Figs. 1-10 : *Hamidia indica* : 1-vegetative hypha, 2-arrangement of oogonia, 3-arrangement of sporangia, 4-arrangement of oogonia, 5-one large egg, 6-emergence of oospore, 7-oogonium with 2 eggs, 8-germinating zoospore within stalked sporangium 9-10-gemmae (Chaudhuri *et al.*, 1947).

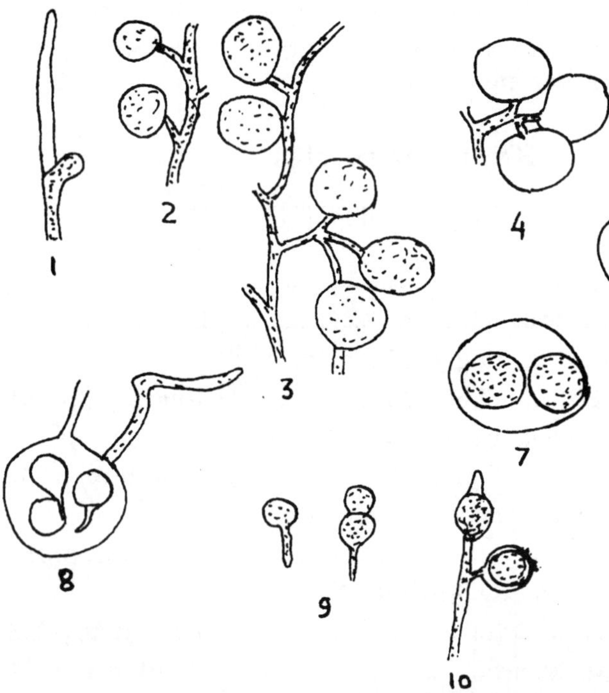

Plate 26

primary zoospore 7.5 to 18 µm in diameter; gemmae few to many, variable in shape, size and position, behave as zoosporangia; oogonia few to many, terminal, lateral, rarely, intercalary; single or in chain, spherical, 30-100 µm in diameter, smooth walled, pitted or unpitted; antheridia absent or present diclinous, monoclinous, androgynous, simple or branched; antheridial cell laterally appressed; oospores one to many, spherical, hyaline, 17.5-40 µm in diameter, centric, subcentric or eccentric, aplerotic.

Six species of *Isoachlya* have been reported from India so far. Their identifying characters are as follows :

Key to the species

Antheridia present .... A

Antheridia absent .... B

A_1 Oospores eccentric C

A_2 Oospores centric or subcentric .... D

B_1 Oogonia abundant, singly or in chain;
Oospores 1-30 is number .... *I. monilifera,* p. 151

B_2 Oogonia occasionally seen; oospores 1-2,
rarely 3 or 4 .... *I. unispora,* p.152

C_1 Oospores one, very rarely two, growth
luxurient .... *I. luxurians,* p. 150

C_2 Oospores 1-18, mostly 3-10, growth
moderate .... *I anisospora,* p.148

D_1 Antheridia diclinous .... *I. toruloides,* p.151

D_2 Antheridia androgynous and diclinous, two
types of spores .... *I. anisospora* var. *indica.,* p.149

Isoachlya anisospora (de Bary) Coker

Bot. Zeit., **46**: 619, 1888

(Plate 27; Figs. 1-3)

Principal hyphae slender, 15-35 µm at base, moderately branched; gemmae abundant, terminal or intercalary, usually spherical or oval, rarely doliform, single or catenulate; sporangia variable in abundance, clavate or filiform, 100-500 × 25-45 µm;

renewed sympodially or by internal proliferation, spores variable in size, smallest 7.5-9 μm in diameter, largest 16.5-18 μm, diplanetic; oogonia abundant, usually lateral, rarely intercalary, spherical or pyriform, doliform when intercalary, 30-100 μm in diameter, predominantly 50-70 μm; oogonial wall smooth, pitted under point of attachment of antheridial cells; antheridial branches usually diclinous, rarely monoclinous, branched; antheridial cells tubular, mostly branched, attached by projections or occassionally laterally appressed to the oogonial wall, oospore eccentric, spherical, 1-18 in number, generally 3-10; 17.5-40 μm in diameter, predominantly 22-30 μm (Srivastava, 1967a).

It was collected from Laldiggi Pond, Gorakhpur on 10th September 1961 for the first time (Srivastava, 1967a).

Isoachlya anisospora (Coker and Matthews) var. *indica* Saksena and Bhargava

Current Science *13*, 79, 1944

Mycelium 8.18-16.36 μm thick, spores of two kinds, smaller 9 μm and bigger ones up to 12 μm in diameter; sporangia 14.5-24.5 μm thick and 99-163.63 μm long; oogonia spherical, terminal, rarely intercalary; wall smooth, 37.14-92.85 μm, mostly 60-66 μm in diameter, wall 1.4 μm thick; antheridia on all oogonia, long, androgynous and diclinous, applied by sides; eggs 1-10 in number, never more than 10, 21.81-55.71 μm in diameter, mostly 24.54-35.45 μm, wall 3 μm thick, aplerotic, centric or subcentric (Saksena and Bhargava, 1944).

This variety was characterized by centric or subcentric oospore and complete absence of eccentric oospore.

This species was isolated on hemp seed baits from a pond, 10 miles from Allahabad (Saksena and Bhargava, 1944) for the first time. Later, it was isolated form waters of Nainital (in summer autumn and spring), Khurpatal (in spring), Bhimtal and Naukuchiyatal (in autumn) lakes when temperature ranged between 15-23.5°C (Khulbe and Bhargava, 1977), pH between 7.4—8 (Khulbe, 1980c); and from clayey soils of the environs of Jalwania pond (pH 8.0-8.5, moisture 30 per cent, organic content 3.5 per cent) of Gorakhpur (Srivastava, 1967b).

Isoachlya luxurians Bhargava and Srivastava

Trans. Brit. Mycol. Soc. *49* : 269-273, 1966
(Plate 27; Fig. 22-25)

Mycelium extensive, diffuse, diameter of 1-week-old colony 3.5-7.5 cm.; principal hyphae slender, thin, 38-52 µm in diam. at the base, narrower towards the tip, moderately branched. Zoosporangia abundant in young cultures, primarily produced on tips of hyphae, filiform to narrowly clavate, straight or curved, 170-835 µm long by 22.5-45 µm in diam.; tip usually straight, sometimes curved and narrowly tapering into an apical discharge tube, occasionally subapical or lateral discharge tubes also occur; secondary zoosporangia usually formed by cymose branching below older ones, internal proliferation also common; spores in several rows discharged through an apical pore or occasionally through a subapical or lateral discharge tube. Zoospores diplanetic, primary zoospores swimming away before encysting, encysted spores 11-13.5 µm in diam.; secondary zoospores discharged from the cysts after an interval and swimming away before re-encysting. Gemmae rare, variable in shape, usually much lobed. Sexual organs appear very early, within 36 hr. after inoculation at 22-25°C; oogonia numerous, borne on slender stalks of variable length; usually lateral, rarely terminal on the hyphae, spherical or slightly subspherical, sometimes pyriform, very rarely cylindrical, with rounded ends in empty sporangia, 40.5-63.5 µm in diam. predominantly 45-58.5 µm, wall moderately pitted, smooth. Antheridial branches usually monoclinous or androgynous where the hyphae are sparse, but diclinous as well as monoclinous where the hyphae are crowded; antheridial cells tubular, clavate or irregular, simple or branched, laterally or sometimes apically appressed to the oogonial wall; fertilization tubes present or lacking, not persistent; oospores usually one, very rarely two in each oogonium; 24.5-56.5 µm diam., mostly between 35-38.5 µm, eccentric, wall 2-3.4 µm thick (Bhargava and Srivastava, 1966).

Collected from Asuran-ka-pokhra Pond, Gorakhpur (U.P), India, 18 September 1961, leg.. G.C. Srivastava.

Type cultures have been deposited in the culture collection of the University of Gorakhpur (India); the University of North Carolina (U.S.A.); at the Herb. Crypt. Indiae Orient. I.A.R.I., New

Delhi (India) and Centraalbureau voor Schimmelcultures, Baarn, Holland.

Kiran and Dayal (1983) reported this species as *Saprolegnia luxurians* (Bhargava and Srivastava) Seymour from Maidagin pond water, Varanasi.

Isoachlya monilifera (de Bary) Kauffman

Amer. J. Bot. *8* : 231, 1921

Proc. Indian Acad. Sci II (2): 137-154, 1935

(Plate 27; Figs. 17-21)

Vegetative growth rather extensive and stout, reaching 1.5 cm on egg yellow in distilled water; hyphae 14-27 µm, occassionally up to 33 µm; zoosporangia not abundant, usually cylindrical to clavate, proliferating internally; zoospores 11.4-12.2. µm; chlamydospores variable, terminal or intercalary, usually in chains, the terminal one being globular or spherical; oogonia very abundant throughout the culture, usually in chains, most of them breaking up easily from the hyphae and each other on maturity, typically spherical with a long neck, wall thin, diameter 49-91 µm, eggs 1-30, mostly 4-12; 21-25 µm, centric; antheridia absent.

Growth in culture-Housefly bait in distilled water showed good growth; zoosporangia developed with zoospores, oogonia not so plentiful. Egg yellow in distilled water and in boiled corn grain-growth luxuriant and plenty of oogonia of normal shape with mature eggs and no antheridia.

This species was collected from Batala and Shalimar, Punjab (Chaudhuri and Kochhar, 1935), which was characterized by constantly naked oogonia and a large number of eggs (1-30), mostly 4-12 in the oogonium. This species was also reported from Jabalpur (Hasija and Khan, 1983; Agarwal *et al.* , 1989) from fresh water of Sardar pond on human nail bait.

Isoachlya toruloides Kauffman and Coker

Amer. J. Bot. 8: 231, 1921

(Plate 27; Figs. 13-16)

Mycelium delicate but vigorous on hemp seeds halves, moderately branched, about 9-36 µm thick; sporangia cylindrical, proliferating internally, abundant in culture; spores diplanetic, 9-14.5 µm in diameter; oogonia occasionally found in the culture, irregular, often seen in empty sporangia, intercalary or terminal, wall thin colourless, pits not visible; antheridial branches diclinous, hyaline, delicate and quickly disappear, antheridia absent on most of the oogonia, usually few, laterally applied, eggs usually 1-8 or up to 16, centric, variable in size, 14-36 µm in diameter; gemmae moderate, irregular.

This species was reported for the first time from India by Dayal (1958). Khulbe (1980) isolated it on hemp seed from Nainital lake.

Isoachlya unispora Coker and Couch

The Saprolegniaceae, p. 85, 1923

(Plate 27; Figs. 4-12)

Mycelium vigorous on hemp seed halves; hyphae irregular, not straight, little branched, about 9-36 µm; sporangia few, primary ones elongated, 360 µm long and 36 µm wide, proliferation very common and internal; spores diplanetic, 9-11 µm at rest, emptying as in *Saprolegnia* and swimming sluggishly, some coming quickly to rest; gemmae plentiful; oogonia occasionally seen in the culture, embedded in a mass of hyphae, mostly spherical, diameter 18-39.6 µm, mostly 30 µm; wall colourless, smooth, generally without pits; eggs usually one to two, rarely 3 or 4, 14.6-39.6 µm thick, mostly 25.2 µm, eggs centric; antheridia never developed (Dayal, 1958).

It was reported for the first time from India by Dayal (1958) from Allahabad.

Plate-27

Figs. 1-3: *Isoachlya anisospora* : 1-Primary empty zoosporangium & lateral secondary zoosporangium; 2-Internal proliferation of primary zoosporangium; 3-Oogonium with variable eccentric oospores & diclinous antheridia.

(Srivastava, 1967a)

Figs. 4-12: *I. unispora* : 4 & 5-Zoosporangia (× 100) 6-8-gemmae (× 100); 9-12-oogonia with oospores (× 400) (Dayal, 1958).

Figs 13-16: *I. toruloides* : 13-15-Zoosporangia (× 100); 16-filamentous oogonium with many centric oospores (× 400) (Dayal, 1958).

Figs. 17-21: *I. monilifera* : 17-Zoosporangium; 18-lateral and intercalary oogonia with many oospores; 19-chain of oogonia with many oospores; 20-oogonium with diclinous antheridium; 21-chain of oogonia without antheridium. (Chaudhuri & Kochhar, 1935).

Figs. 22-25: *I. luxurians:* 22-Mycelium with primary apical zoosporangium, lateral secondary zoosporangium, and lateral oogonia with androgynous antheridia (Bhargava & Srivastava, 1966); 23-Hypha with lateral zoosporangia. 24-Gemmae; 25-lateral oogonia with androgynous antheridia.

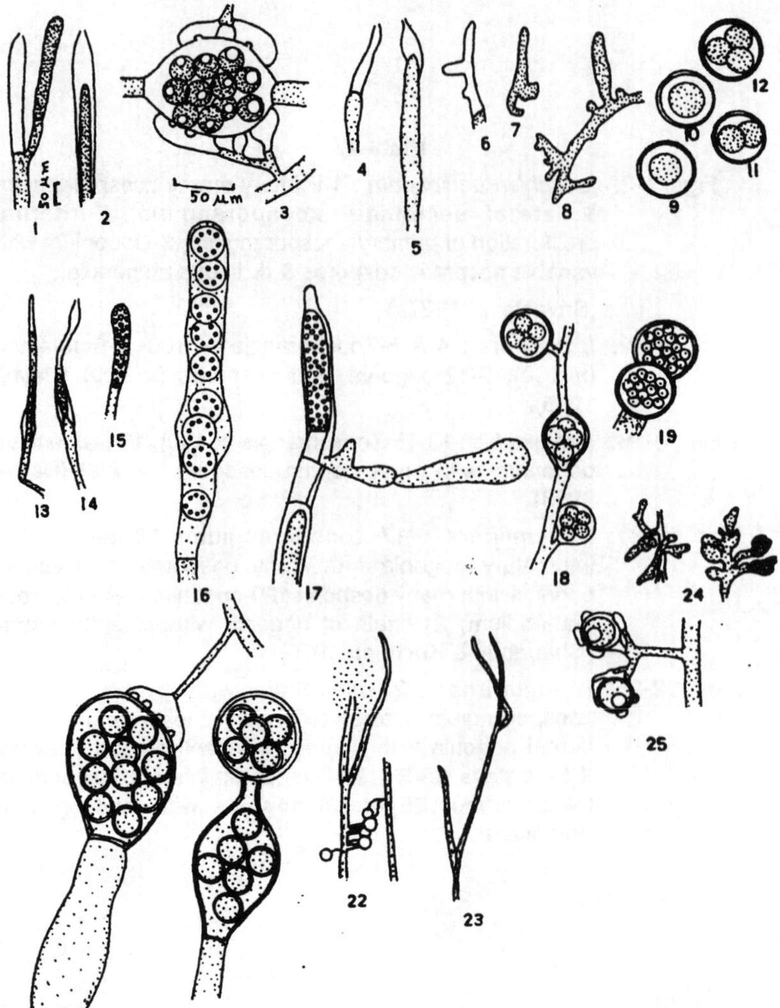

Plate 27

Leptolegnia de Bary

Bot. Zeit. **46** : 631, 1888.

The Saprolegniaceae, p. 157, 1923

Two species of *Leptolegnia* have been reported from India so far.

Key to the species

1. Oogonia smooth walled, antheridia mostly diclinous, rarely monoclinous, oospores not maturing .... *L. caudata*, p.155
2. Oospores with irregular projections, antheridia androgynous, oospores eccentric...... *L. eccentrica*, p.158

Leptolegnia caudata de Bary

Bot. Zeit., **46** : 631, 1888

(Plate 28; Figs. 1-7)

Principal hyphae slender, 3-12 µm diameter at the base, mostly 10-11.5 µm; primary zoosporangia abundant, 350-1250 µm long and 12-20 µm in diameter, predominantly 750 × 15 µm; secondary zoosporangia internal, rarely sympodial; zoospores arranged in a single row, 15-45 µm long and 7.5-15 µm in diameter, predominantly 20-30 × 8-12 µm; encysted primary zoospores 11-15 µm in diameter; gemmae absent; oogonia abundant, lateral, formed cymosely on short to long stalks, developed at the distal part of the hyphae, spherical, subspherical, rarely irregular in shape, 25-45.5 µm in diameter, mostly 30-35.5 µm; oogonial wall smooth and unpitted; antheridia diclinous, rarely monoclinous, 1-4 per oogonium, 8-18.5 µm thick; oospheres not maturing (Khulbe, 1985).

Isolated on hemp seed from Nal-Damyanti Tal and Sat Tal on March 15, 1980, leg. R.D. Khulbe. The type culture has been deposited in the Department of Botany, Kumaun University, Nainital.

Plate-28

Figs. 1-7 : *Leptolegnia caudata* : 1, Showing arrangement of zoospores in single row; 2, liberation of zoospores; 3, internal proliferation; 4, *in situ* germination of zoospores; 5, young oogonium with monoclinous antheridium; 6, young oogonium with diclinous antheridium; 7, mature oogonium attended by three antheridia showing disorganising oogonial contents.

Scale A: Figs. 1-4; Scale B: Figs. 5-7. (Khulbe, 1985)

Fig. 8 : *L. eccentrica* : Lateral oogonia with spiny walls, single eccentric oospore and androgynous antheridium.

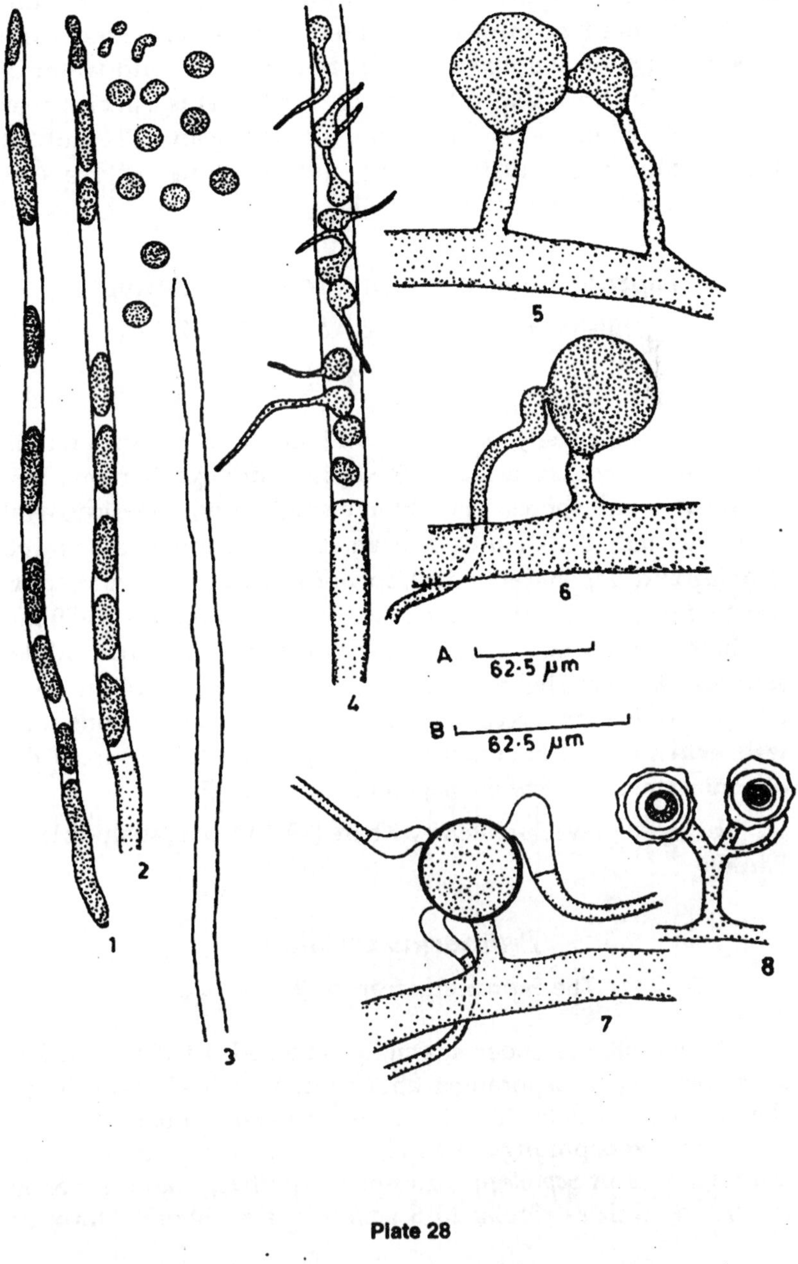

Plate 28

This species was isolated first of all from water of Nal Damyanti Tal in March 1974 (RDK Slide No. L/1), but could not be cultured in pure form. Later, it was reisolated from infected eggs of *Cyprinus carpio* of Bhimtal lake and water of Ram tal (Sat Tal) lake (Mer *et al.,* 1981). This species was again isolated on hemp seeds from Nal Damyanti Tal and Sat Tal on March 15, 1980 (Khulbe, 1985). Type culture was deposited at Nainital (RDK CC-L/1).

Leptolegnia eccentrica (de Bary) Coker

J. Elisha Mitchell Sci. Soc. **42** : 207-226, 1927

(Plate 26; Fig. 8)

Hyphae sparingly branched, 5-12 µm thick; zoosporangia long, filamentous, same size as the hyphae or slightly more, 385-980 by 9-19 µm; spores typically in a single row, spherical, oval or elongated, 8-18.5 µm in diameter; zoospore discharge leptolegnoid; oogonia abundant, spherical to oval with numerous short irregular projections, including these projections 16-29 µm in diameter, borne terminally or laterally; oospores single, spherical to oval slightly irregular, filling the oogonium, 13.5-22.8 µm in diameter, eccentric with a large oil drop on one side, wall extremely thick; antheridia androgynous arising from the oogonial stalk or below the oogonium.

This species was isolated by Khare (1992) from Jabalpur lake, Bareilly.

Protoachlya Coker

The Saprolegniaceae, p.. 90, 1923.

Hyphae stout, slender, sparingly branched, 18-45 µm thick at base; primary zoosporangia abundant, terminal, cylindrical, clavate or naviculate; 15-52.5 µm in diameter, 95-600 µm long; secondary zoosporangia lateral, cymose, sympodial, rarely intercalary as in *Saprolegnia;* zoospores discharge apically as in *Achlya* as well as similar to *Saprolegnia;* also liberate through lateral papillae; zoospores diplanetic, 7-11 μm in diameter when encysted; gemmae many, variable in shape, size and position; oogonia many, lateral on main hyphae, intercalary, spherical, pyriform, oogonial wall smooth or papillate, pits rarely present; antheridial branches diclinous, androgynous, monoclinous; oospores 1-10, 10-14 μm in diameter, centric, subcentric and eccentric.

Two species of *Protoachlya* have been reported from India. Their differentiating characters are as follows :

Key to the species

1. Oogonia strictly smooth walled, antheridia mostly diclinous, rarely androgynous, oospores centric and/or subcentric ....*P. paradoxa*, p. 162
2. Oogonia smooth walled and papillate, antheridia mostly androgynous, rarely diclinous and monoclinous, oospore eccentric .... *P. oryzae*, p. 159

Protoachlya oryzae Khulbe

Current Science **52**(9): 428, 1983

(Plate 29; Figs. 1-17)

Vegetative growth dense on hemp seed halves, 18-40 μm in diam. at base; zoosporangia many, primary zoosporangia formed at the tip of hyphae, and secondary ones usually formed by cymose branching, below older ones, many papillate; long, cylindrical, 22-52.5 × 95-600 μm; zoospores liberate through apical pore or subapical or lateral papillae, diplanetic, 8-10 μm in diam., gemmae many; oogonia many, spherical, wall smooth or papillate, 30-65 μm in diam., papillae 5-13 μm in diam., antheridia androgynous, rarely diclinous or monoclinous; oospore single, rarely 2-3, 18-38 μm in diam., mostly 22.5 μm and eccentric (Khulbe, 1983a, 1985).

This species was isolated from infected roots of *Oryza sativa* from Bhimtal (Nainital) on 18th July, 1981 (Khulbe, 1983a, and from *Fagopyrum esculentum* from Russey Village near Nainital (Khulbe, 1985).

Plate-29

Figs. 1-17 : *Protoachlya oryzae* Khulbe

1. Single hyphal tip.
2. Young zoosporangium..

3&4. Zoosporangia showing liberation of zoospores and a few encysted zoospores at the base.

5-10. Mature zoosporangia showing zoospores liberation through apical pore·as wall as lateral papillae.

11 Oogonia with diclinous antheridia.

12. Lateral smooth-walled oogonium without antheridium.
13. Lateral papillate oogonium.
14. Lateral oogonium with androgynous antheridia.
15. Oogonium with single outgrowth.
16. Long coiled oogonial stalk.
17. Oogonium with short stalk and many long papillae.

Figs. 1-6, 8-12 & 14-16, scale A; Figs. 7&17, scale B; Fig. 13, scale C.

(Khulbe, 1985)

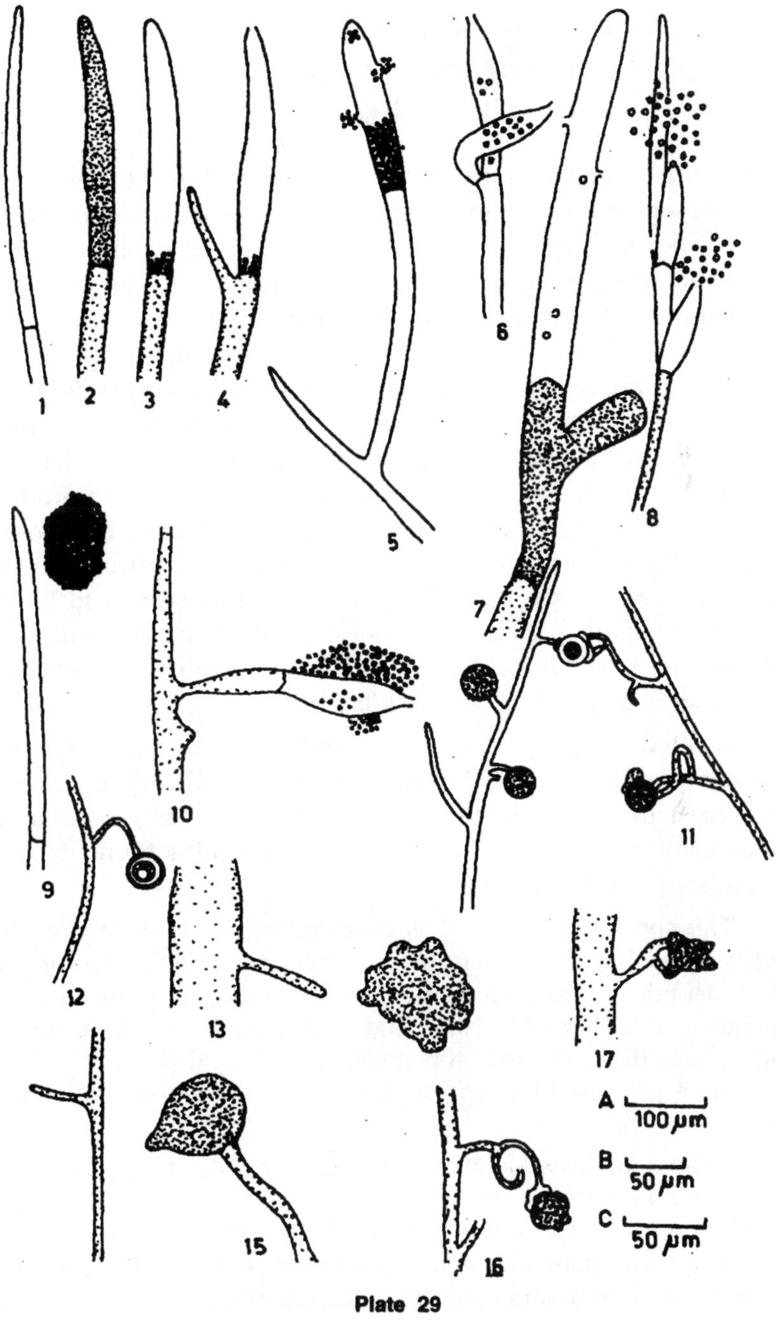

Plate 29

Protoachlya paradoxa **Coker**

The Saprolegniceae, p. 91, pl. 26-28, 1923.

(Plate 30; Figs. 1-11)

Principal hyphae stout with blunt tips, 18-45 µm thick at the base; zoosporangia abundant, cylindrical, clavate or naviculate, 200-500 µm long by 15-50 µm in diameter, mostly 280 × 25 µm; secondary zoosporangia lateral in cymose manner or sympodially arranged, rarely internal as in *Saprolegnia*; zoospores discharge apically as in *Achlya*, rarely as in *Saprolegnia*; zoospores diplanetic, variable in behaviour, liberate like *Achlya* and *Saprolegnia*, 7-11 µm in diameter; gemmae elongated or flask shaped, sometimes 2-8 in a chain, bahave as zoosporangia; oogonia abundant, lateral on main hyphae, sometimes intercalary, spherical or pyriform, 30-78 µm in diameter, mostly 62.5 µm; oogonial wall smooth, pits rarely present; antheridial branches diclinous, rarely androgynous, 1-3 per oogonium; antheridial cell short to long, clavate or with tuberculated projections; oospores 1-10 per oogonium, mostly 3-4, spherical, centric or subcentric, 10-40 µm in diameter, generally 26 µm (Khulbe, 1985).

Isolated on hemp seed and house-flies from forest soils of Nainital on March 17, 1980, leg. R.D. Khulbe. The type culture has been deposited in the Department of Botany, Kumaun University, Nainital (PR/2). This species was also found in the forest soils of Kilbury and Ramgarh.

This species was isolated as a saprophyte on house fly, egg's white from Lahore (Chaudhuri and Kochhar, 1935; Chaudhuri and Banerjee, 1942); on hemp seeds, Maidagin pond water, Varanasi in March 1977 (Kiran and Dayal, 1983); on hemp seeds and house flies from oak forest soils of Nainital (Khulbe, 1985) and as a parasite of *Cyprinus carpio* var. *specularis* (Sati and Khulbe, 1980).

Kiran and Dayal (1983) reported variable zoosporangial behaviour, i.e., some of the spores swim away slowly on emerging, while the other remaining attached to the sporangium mouth and encysting there. Gupta and Mehrotra (1989a & b) reported its occurrence in Brahma sarovar of Kurukshetra.

Plate-30

Figs.1-11: *Protoachlya paradoxa* Coker.

1. Young hyphal tip.
2. Young zoosporangium with a lateral papilla.
3. Mature zoosporangium.
4. Liberation of zoospores through apical and lateral papillae.

5&6. Gemmae.

7. Zoosporangia showing achlyoid and saprolegnoid zoospores release.
8. Chain of gemmae.
9. Zoosporangium showing *in situ* germination of zoospores.
10. Young oogonium with two diclinous of antheridia.
11. Mature oogonium showing four centric oospores and attachment of two diclinous antheridia.

Figs. 1-7, scale A; Figs. 8-10, scale B; Fig. 11, scale C.

(Khulbe, 1985)

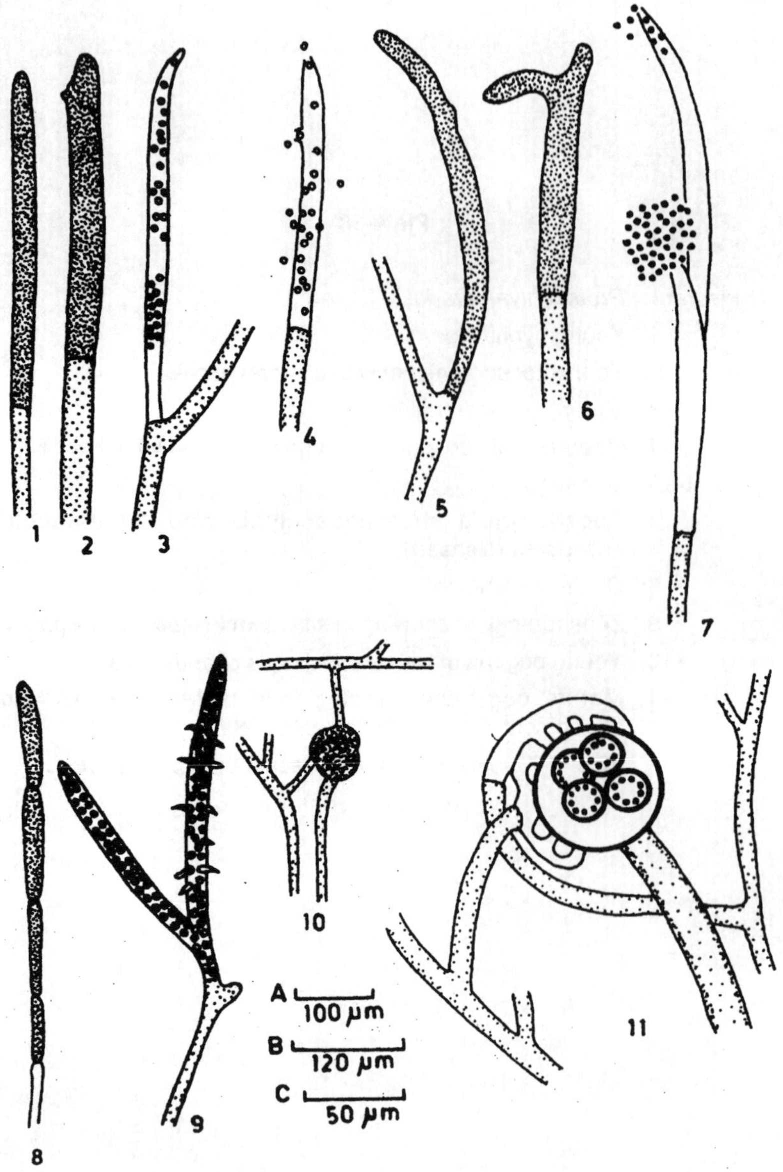

Plate 30

Pythiopsis de Bary

Bot. Zeit. **46**: 597, 613, 629 & 645, 1888

· The Saprolegniaceae, p. 17, 1923

Hyphae slender, much or little branched; sporangia typically short and plump, spherical, oval, pyriform with a distinct apical papilla, or varying to elongated and irregular, primarily borne at the tips of the hyphae and multiplied from lateral stalks below the old ones to form more or less dense clusters; spores emerging and swimming as in *Saprolegnia*, pip shaped with two apical cilia, sprouting after the first encystment (monoplanetic); gemmae resembling the sporangia or oogonia, formed plentifully, often in chains, producing zoospores after a rest; oogonia borne like the sporangia and gemmae and resembling them in youth, typically spherical, oval or pyriform with unpitted walls, smooth or with a few blunt papillae; antheridia short and thick, typically androgynous from the close neighbourhood of the oogonia; eggs one or few, eccentric, with a lunate cap of droplets on one side in *P. cymosa* (Coker, 1923).

In India, two species, viz., *P. intermedia* and *P. humphreyana* have been isolated. The key to these species is as under :

Sporangia mainly globular or clavate .... A

A_1 Secondary zoosporangia internal as well as cymose, lateral; egg single, 22.4-28.8 µm in diameter; antheridium androgynous .... *P. intermedia*, p.165

A_2 Secondary zoosporangia strictly cymose; egg 1-2, 30-37.5 µm in diameter; antheridium just below the oogonium, ... *P. humphreyana*, p. 166

Pythiopsis intermedia Chaudhuri and Banerjee

Proc. Indian Acad. Sci. **15** : (4)B, 1942, pp. 216-224

(Plate 31; Figs. 1-10)

Hyphae slender, 4.8-5.5 µm in diameter at base, much branched; zoosporangia globular or clavate; usually proliferating internally and borne in a cymose manner with two apical papillae; secondary zoosporangium grow partly inside and partly outside the empty primary zoosporangium; in cymose branching the secondary zoosporangium formed either very near the primary zoosporantium or at a fairly long distance away; zoospores usually 9.5 µm, biciliate, monoplanetic; oogonia plentifully formed in old cultures, spherical, unpitted, 25.6-35.2 µm in diameter, eggs 22.4-28.8 µm in diam., single, eccentric; antheridial branches long arising far away from the oogonium, one to each oogonium, androgynous, clavate, later curving along the oogonial wall; gemmae resembling zoosporangia formed at the tips of the hyphae but have a long drawn out apex, gemmae formed if the mould is kept in stagnant water for a long time or if the acid in the medium is slightly increased.

Collected from a soil sample taken from a drain in Lahore in November, 1938 (Chaudhuri and Banerjee, 1942).

In this species zoosporangia were mostly globular (32-41.6 µm), some were elongated (31.2 -23.2 µm or 28.8 × 22.4 µm). The hyphae were much thinner being 4.8-5.5 µm, while in the other species (*P. humphreyana*) this thickness is much greater. Most important point is the presence of internal proliferation, not to be found in *P. cymosum* or *P. humphreyana*. Zoospores, oogonia and eggs were more or less intermediate in size between 2 species.

Pythiopsis humphreyana Coker

Mycologia 6: 292,1914

The Saprolegniaceae, p. 20, 1923

Mycelium diffused; principal hyphae slender, branched, 12-15 µm in diam., thicker at the base of the reproductive bodies; sporangia spherical, oval or pyriform, 37.5-60 µm in diam., renewed in cymose manner; zoospores discharge through apical papilla, monoplanetic; oogonia borne exactly like the sporangia, usually spherical, rarely irregular, 37-50 µm in diam., wall smooth, antheridia present, usually develop just below of the oogonia, rarely diclinous; oospheres usually one, very rarely two, spherical and filling and oogonium, 30-37.5 µm in diameter.

It was collected from clayey soils of the environs of Jalwania Pond (pH 8.5, moisture content 30 per cent, organic matter 4.0 per cent) (Prabhuji and Srivastava, 1978 a&b).

Plate-31

Figs. 1-10 : *Pythiopsis intermedia*

1. Internally proliferating zoosporangium.
2. Zoosporangium with two apical papillae.
3. Cymosely borne zoosporangia.
4. Internal proliferation of zoosporangium.
5. Abnormal internal proliferation.
6. Knob formation in internal proliferation.
7. Zoosporangium half a minute before the liberation of zoospore mass.
8. Zoospores mass coming out in semidifferentiated state.
9. Hypha with two oogonia.
10. Oogonium with androgynous antheridium.

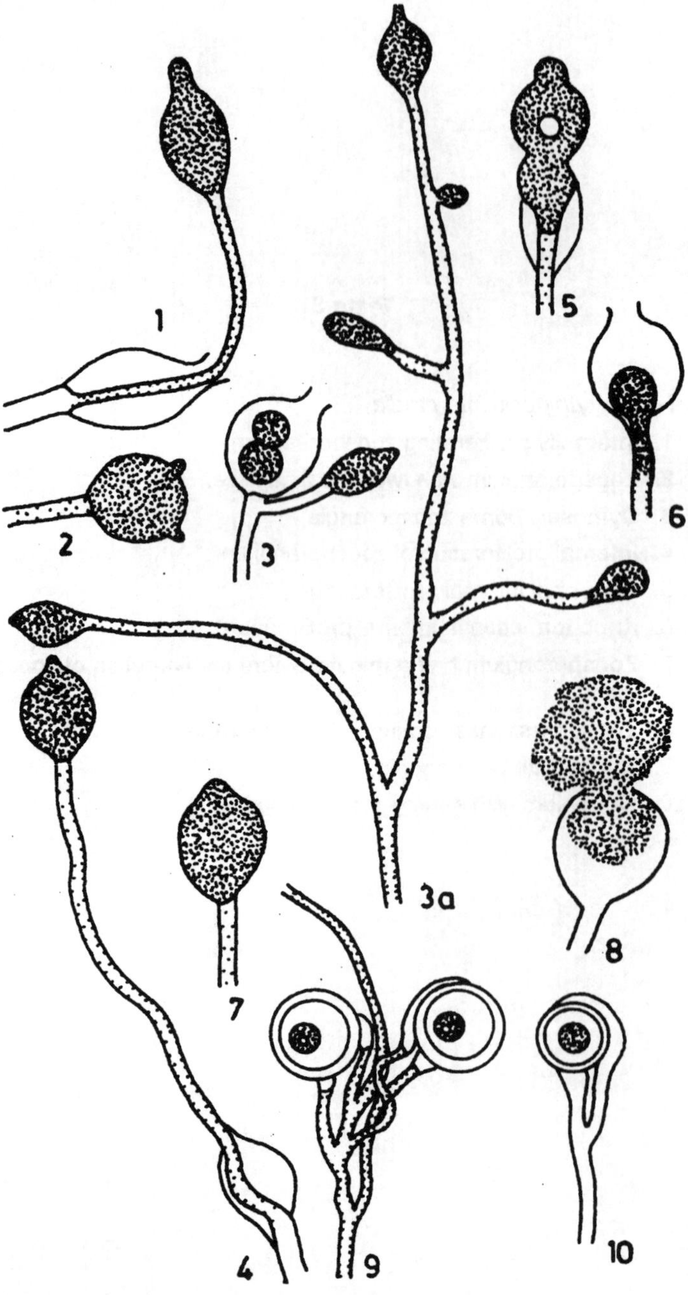

Plate 31

Saprolegnia Nees v. Esenbeck

Nova Acta Acad. Leop. Carol. ***11*** : 513, 1823

The Saprolegniaceae p. 22, 1923.

Aquatic Phycomycetes, p. 835, 1960.

Nova Hed. Krpto. **19** : 1-124, 1970

Key to the species

	Sex organs present	A
	Sex organs absent	B
A_1	Antheridia present in all or some oogonia	C
A_2	Antheridia completely absent	D
B_1	Gemmae persistent	*S. parasitica*, p.182
B_2	Gemmae deciduous	*S. parasitica* var. *kochhari*, p.183
C_1	Antheridia present in all oogonia	G
C_2	Antheridia not present in all oogonia	H
D_1	Oospores eccentric	*S. eccentrica*, p. 174
D_2	Oospores centric	E
D_3	Oospores centric and subcentric	*S. unispora*, p. 186
E_1	Oospores variable in shape	*S. irregularis*, p.178
E_2	Oospores spherical	F
F_1	Oospores 1-16, mostly 2-5, oogonia catenulate	*S. monilifera*, p.181
F_2	Oospores 9-12, mostly 12, oogonia not catenulate	*S. rhaetica*, p.184
F_3	Oospores many	*S. lapponica*, p.179
G_1	Antheridial branches present	J
G_2	Antheridial branches absent, antheridial cells hypogynous	*S. hypogyna*, p.177
H_1	Antheridia present on more than 50 per cent, oogonia, oospores 1-25, mostly 6-10 in number	*S. mixta* var. *asplundii*, p.180
H_2	Antheridia present on less than 50 per cent oogonia	I

I_1 Antheridial branches diclinous, oospores 10-40, mostly 15-20 per oogonium ... *S. crustosia* var. III, p.172

I_2 Antheridia mostly androgynous, rarely diclinous, oospores 1-40, mostly 6-18 per oogonium .... *S. ferax*, p.175

J_1 Antheridial branches glomerate (extensively branched) .... *S . glomerata*, p.176

J_2 Antheridial branches not glomerate .... K

K_1 Antheridial branches strictly diclinous or mostly diclinous, rarely androgynous L

K_2 Antheridial branches androgynous or monoclinous or both .... M

L_1 Oospores centric .... N

L_2 Oospores eccentric or subcentric .... *S. anisospora*, p.171

M_1 Oospores centric and subcentric .... O

N_1 No internal proliferation in primary sporangia .... *S. bhargavii*, p.172

N_2 Secondary zoosporangia internal as well as lateral .... P

O_1 Oogonial wall variable, smooth and papillate .... *S. subterranea*, p.184

O_2 Oogonial wall smooth .... R

P_1 Gemmae few, oospores 4-20, mostly 8-10 ... *S. kauffmaniana*, p. 178

P_2 Gemmae abundant .... Q

Q_1 Oospores 1-16, mostly 30, size, 22-29 µm .... *S. delica*, p. 173

Q_2 Oospores 1-40, mostly 4-18, size 8-44 µm, mostly 20-23.5 µm .... *S. diclina*, p. 173

R_1 Oospores strictly centric .... S

R_2 Oospores centric and subcentic both .... T

S_1 Oospores 1-30, antheridia simple .... *S. litoralis*, p.179

S_2 Oospores 1-15, antheridia bent .... *S. monoica*, p.181

T_1 Oogonia smooth, antheridia mostly monoclinous, rarely diclinous .... *S. uliginosa*, p.185

T2 Oogonia smooth or with sparse papillation, oogonia and oospores yellowish, antheridia androgynous from just below the oogonium....*S. terrestris*, p.185

Saprolegnia anisospora de Bary

Bot. Zeit. **46** : 619, 1888

Syn. *Isoachlya anisospora* (de Bary) Coker and Matthews

North Amer. Flora 2 : *26* : 1937

(Plate 38; Fig. 15)

Mycelium profusely branched, coenocytic, hyaline, moderately thick, usually delicate; thickness varies from 16-33 µm in diameter; primary sporangia elongated and cylindrical, 13-33 µm in width and 66-330 µm in length; two types of sporangia observed, in one the entire protoplast is segmented and transformed into zoospores (dictyoid sporangia), whereas in second type sporangia unseptate and typical saprolegnoid types containing spherical encysted zoospores, which are variable in size, usually 6.6-13.2 µm in diameter (anisosporous); gemmae in abundance, greatly varied in shape and size, terminal and often catenulate, the cells are in long chain, the lower cells are more elongated than the upper cells, they are more or less rectangular or subspherical, later on behaving as zoosporangia, rarely intercalary; oogonia in abundance, mostly terminal, rarely intercalary; oogonia in abundance, most terminal, rarerly intercalar or lateral, spherical, smooth walled and unpitted, varies from 49.5-92.5 µm in diameter; oogonial stalk short, stout, straight; oospores thick or thin walled, eccentric or sub-centric, filling the oogonium, varying in number from 1-20, size 19.8-23.1 µm in diameter; antheridia diclinous and tuberous; the germination of oospores and fertilization tube not observed.

Isolated on mustard seed from Engineering College lake water, Jabalpur, M.P., Nov., 1978, at water temp. 24°C and pH 7.6, leg. Khan, 1981 (Hasija and Khan, 1983), specimen HCIO, No. 33745-6; pond waters of Lahurabir, Varanasi, April, 1978 (Kiran and Dayal, 1983); Robertson lake, Jabalpur, post monsoon season (Aggarwal *et al.*, 1989); Supa Tal lake, Jabalpur (Indira *et al.*, 1988); surface water and on bottom sediments of polluted Ganga Sagar lake, Jabalpur, October, January, February, March to June, August and September (Mishra *et al.*, 1990); from Jalalpur lake and Dhopeshwar pond, Bareilly (Khare, 1992).

Saprolegnia bhargavii Khulbe and Verma

Nova Hedwigia **44** (1-2): 215-219, 1987

(Plate 32; Figs. 1-3)

Mycelium branched, 15-30 μm in diameter at the base; zoosporangia long, filiform, 105-678 μm long by 18.15-29.5 μm in diam.; secondary zoosporangia formed by cymose or basipetal branching, internal proliferation completely absent; gemmae abundant, spherical, single, sometimes in chain; oogonia numerous, mainly lateral, 37-72.5 μm in diameter, wall smooth and pitted; antheridia usually diclinous; oospores centric, 1-7 per oogonium, usually 3-5, 18-28.5 μm in diameter, predominantly 22.5-25 μm.

Isolated from cultivated soil of Khatima, October and November, 1980. Type culture at Botany Department, and C.M.I. Kew, England, IMI-277414 (Khulbe and Verma, 1987).

Saprolegnia crustosa var. III Maurizio

Mitt. d. Deutsch. Fischerei-Vereins 7 : 52, 1899

(Plate 36; Figs. 3-7)

Principal hyphae slender, 10-37.5 μm thick at base, moderately branched; sporangia abundant; variable in shape, ranging from fusiform, cylindrical, filiform to naviculate; 87-390 μm long and 23.2-58.0 μm in diameter; renewed by internal proliferation; spores 10-11.5 μm in diameter, diplanetic; gemmae abundant, variable in shape and size; oogonia abundant, mostly lateral, rarely terminal, occasionally formed in empty sporangia, mostly spherical, rarely ovoid or pyriform, sometimes cylindrical, 52-93 μm in diameter; oogonial wall smooth, thin, with few to many conspicuous pits; antheridial branches diclinous, present only on about one third of the oogonia; antheridial cells tubular, simple or branched; laterally appressed or rarely attached by projection to the oogonial wall; oospore centric, spherical, 10-40 in number, mostly 15-20, 21.5-27 μm in diameter, predominantly 23-25 μm (Khulbe, 1980).

Isolated on maize seed bait from Bhim Tal lake on 5th April, 1975 for the first time (Khulbe and Bhargava, 1977), and found to be distributed in lake waters of Nainital (in summer, rainy and spring), Sariya Tal (in all the seasons except spring), Khurpa Tal and Bhim Tal (in spring) and Naukuchiya Tal (during winter and spring) when temperature ranged from 15-23..5°C (Khulbe and Bhargava, 1977; Khulbe, 1980b) and pH between 7.5-8.2 (Khulbe, 1980b).

Saprolegnia delica Coker

The Saprolegniaceae, p. 30, 1923

(Plate 32; Figs. 4-6)

Principal hyphae moderately stout, branched, 13.5-49.5 µm thick at base; sporangia abundant; cylindrical; 150-450 µm long by 15-68 µm in diameter, predominantly 20-150 µm; renewed internally; spores 10-11 µm thick in diameter, mostly 10 µm; gemmae abundant; shape variable from spherical, fusiform to clavate, single or in chain of 2-6; mostly terminal, sometimes intercalary; oogonia abundant; spherial or oblong, mostly terminal on main or lateral hyphae; 33-115 µm in diameter, predominantly 65-85.5 µm; oogonial wall smooth, thin with few conspicuous pits; antheridial branches abundant, diclinous, rarely androgynous; simple or branched, 2-8 per oogonium; antheridial cells tuberous; oospores centric; spherical; 1-30 in number, mostly 4-8; dark coloured; 16.5-29 µm in diameter, predominantly 27 µm (Khulbe, 1980).

Isolated from some temporary ponds (pH 7.5-8) of Varanasi during September (Dayal and Thakurji, 1966); on maize seed from Bhim Tal lake, 5 April 1975 (Khulbe and Bhargava, 1977; Khulbe, 1980b), and from soil and water samples around I.A.R.I., New Delhi during winter (November to February) when temperature ranged between 10-25°C (Chowdhry and Aggarwal, 1980 a&b). Khare (1992) isolated it from Jalalpur lake and Dhopeshwar pond, Bareilly.

Saprolegnia diclina Humphrey

Trans. Amer. Phil. Soc., 17 : 109, 1892 (1893)

(Plate 35; Figs. 1-3)

Principal hyphae slender, branched, 12-68 µm thick at base; zoosporangia abundant, cylindrical, rarely filamentous, 45-650 μm long by 25-85 μm in diameter, predominantly 120-150 × 28.5-30 μm, renewed internally; encysted zoospores 9.5-11 μm in diameter, predominantly 10.0-12 μm; gemmae abundant, spherial or cylindrical, single or in chain or 2-4, terminal or intercalary; oogonia abundant, terminal or lateral on short to long stalks, spherical or oval, rarely pyriform, 30-179.0 μm thick in diameter, predominantly 50-70 μm; oogonial wall smooth, thin, with few inconspicuous pits; antheridial branches diclinous, simple or branched, slender, 2-6 per oogonium; antheridial cells tuberous; oospores spherical, centric. 1-40 in number, mostly 4-18, 18-44 μm in diameter, predominantly 20-23.5 μm (Khulbe, 1980).

This species was isolated on hemp seeds from Sarnath, Varanasi (Thakurji, 1967); Ramgarh lake, Gorakhpur (Srivastava, 1967); soil and water samples around I.A.R.I., New Delhi during winter from November to February at 10-25°C temperature (Chowdhry and Agarwal, 1980 a&b); some lakes of Jabalpur containing sufficient organic matter (26.3 ppm) and chloride (166.3 ppm). Supa Tal lake, Jabalpur (Indira *et al.*, 1988); Robertson ponds, Jabalpur during summer (March to June), monsoon (July-September), post monsoon (July-September) and Hanuman Tal pond during winter (December-February) and summer of 1978-1979 (Agarwal *et al.*, 1989); from Brahma Sarovar, Kurukshetra (Gupta and Mehrotra, 1988a); polluted water of Ganga Sagar, Jabalpur during February to July, 1988 (Mishra *et al.*, 1990); river, pond and lakes of Bareilly (Khare, 1992), under subtropical and temperate conditions of Kumaun at 0.8-4.9 per cent organic carbon (Khulbe, 1985 & 1991).

Saprolegnia eccentrica (Coker) Seymour

The Genus *Saprolegnia*, p. 53, 1970

(Plate 38; Figs. 18 & 19)

Hyphae slender, delicate, sparingly branched, 13.8-58.17 μm in diameter, gemmae abundant, irregular, functioning as sporangia; sporangia abundant, filiform, 240.9-338.8 × 27.27-38.88 μm, renewed only by internal proliferation; zoospores discharge saprolegnioid, rarely dictyoid; encysted spores 5.5 to 11.08 μm in diameter; oogonia abundant, lateral or terminal, spherical, occasionally pyriform, 27.7-41.55 μm in diameter; oogonial wall smooth, unpitted; oogonial stalk stout and straight, sometimes ellipsoidal, almost filling the oogonium, 22.1-33.2 µm in diameter, eccentric.

Isolated as a saprophyte from pond waters of Sarpat Ganj, Lucknow (Mishra, 1980 & 1982), Pune in December (Kapadnis, 1980), and from Brahma, and Sanhit Sarovar, Kurukshetra (Gupta and Mehrotra, 1988a).

Saprolegnia ferax (Gruith) Thuret

Ann. Sci. Nat. Bot. III, ***14*** : 214, 1850

(Plate 35; Figs. 4-7)

Principal hyphae moderately stout, branched, 7.5-80 µm thick at base; zoosporangia abundant, cylindrical or filamentous sometimes tapering, 80-650 µm long by 7.5-67 µm in diameter, predominantly 180-350 × 15-35 µm, renewed internally; encysted zoospores 9-12 µm in diameter, predominantly 10 µm; gemmae few, spherical, pyriform, cylindrical, irregular, terminal intercalary; oogonia abundant, spherical or cylindrical, sometimes filamentous, terminal on main hyphae or lateral on short to long stalks, sometimes intercalary, very rarely in the old empty zoosporangium, 30-200 µm in diameter, predominantly 60 µm; oogonial wall smooth, thin with numerous conspicuous pits, antheridial branches mostly androgynous, sometimes diclinous, less than 20% of the oogonia have antheridia; antheridial cells tuberous, 1-4 per oogonium, laterally appressed; oospores spherical, centric, 1-40 in number, mostly 6-18, 12-44 µm thick in diameter, predominantly 19.8-28 µm (Khulbe, 1985).

This species was isolated on hemp seeds from some ponds of Allahabad for the first time by Dayal and Tandon (1962). Later, it was reported from different places under various ecological conditions, viz., Asuran-ka-Pokhra Pond, Gorakhpur (Srivastava, 1967a); permanent, temporary and cemented ponds of Varanasi during December to April at pH 7-8.5 (Dayal andThakurji, 1966 and 1968b); ponds of Varanasi (Thakurji, 1967); soil and water samples around I.A.R.I., New Delhi during winter (November to February) at 10-25°C temperature (Chowdhry and Agrawal, 1980 a&b); alkaline ponds of Lucknow (Misra, 1982), soils of Gorakhpur (Prabhuji, 1984), tomato field soils at Nanakmatta on April 17, 1982 (Verma, 1984), and from Brahma and Sanhit Sarovar (Gupta and Mehrotra, 1988a).

In Kumaun Himalayan colder region, this species was isolated on hemp seeds from Nainital lake on March 8, 1973 (Khulbe and Bhargava, 1977) and found to be distributed in other lakes (Sat tal, Khurpa Tal, Bhim Tal and Naukuchiya Tal) during all the seasons at 10-28.5°C water temperature (Khulbe and Bhargava, 1977), 7.3-8.8 pH (Khulbe, 1980c) and a wide range of nitrate, sulphates and phosphates (Khulbe and Bhargava, 1983). It was also collected from soils of Tiffin Top, Government House, Nainital (Sati and Khulbe, 1980); rhizosphere and non rhizosphere soils of tomato, paddy, pea and wheat fields around Nainital (Khulbe, 1983b) and Ram Tal waters (Mer and Khulbe, 1984).

The occurrence of this species was also found to extend from aquatic habitats (ponds, tanks, streams, rivers and lakes) to certain terrestrial cultivated (Almora, Dwarahat, Ganai, Ranikhet, Haldwani, Ramnagar, Berinag, Lohaghat and Pithoragarh) and non cultivated and forest soils (Ranikhet, Haldwani, Jyolikot, Kilbury, Naina Peak and Nainital), and appeared frequently in oak forest soils of Naina Peak, Nainital; and sal forest soils of Haldwani during all the seasons at 5-30°C soil temperature, 5.3-6.9 pH, 2-90 per cent moisture, and 0.4-4.9 per cent organic carbon (Khulbe, 1985 and 1991). Khare (1992) isolated it from all the rivers, pond and lakes of Bareilly.

Saprolegnia glomerata (Tiesenhousen) Lund

Kgl. Danske. Selsk. Skrift., Naturv. Math. Afd. IX **6** (1): 14, 1934

(Plate 35; Figs. 8-9)

Hyphae moderately stout, branched, 15-65 µm in diameter at base; zoosporangia abundant, filiform, clavate, straight, rarely bent, 150-550 µm long by 30-75 µm, renewed internally; zoospore discharge saprolegnoid, encysted zoospores 10-15 µm in diameter, mostly 11.5 µm; gemmae few, clavate, spherical, rod shaped, terminal, single or catenulate, size variable, mostly 120-180 µm long and 15-50 µm in diameter, function as zoosporangia; oogonia abundant, terminal , lateral, spherical, pyriform, rarely cylindrical, 45-150 µm in diameter, mostly 65.5 µm, cylindrical oogonia 105-315 µm long and 26-30 µm in diameter; oogonial wall smooth, conspicuously pitted; oogonial stalks variable in length, generally 1-2 times longer than the diameter of the oogonium, stout, straight, sometimes bent, simple or branched; antheridia on all oogonia, androgynous, rarely monoclinous, abundant, simple, branched, irregular, compactly-clustered (glomerate); oospores spherical, centric, 1-15 in number, generally 5-8, 18-32 µm in diameter, mostly 25.5-26.2 µm.

This specie was isolated on hemp seeds from Ram Tal soils (Sat Tal, Nainital) in March, 1979 for the first time (Mer and Khulbe, 1984). Later, it was found to be distributed in some cultivated field soils of Sat Tal, Dwarahat and Ranikhet in July, 1980 (Khulbe, 1985), in oak forest soils of Naina Peak, Nainital during spring and summer, and in forest soil of Haldwani in all the seasons at temperature 5-30°C, pH 5.3-6.9, moisture 2-90 per cent and organic carbon 0.4-4.9 per cent (Khulbe, 1991).

Saprolegnia hypogyna (Pringsheim) de Bary

Jahrb. f. Wiss. Bot. **9**: 191, 1873,

Bot. Zeit., **41**: 56, 1883

The Saprolegniaceae, p. 60, 1923.

(Plate 38; Fig. 20)

Hyphae slender, branched, 15-68 µm in diameter; zoosporangia abundant, filiform or clavate, single, rarely curved, irregular, 85-500 µm long and 15-35 µm thick, mostly 120-150 × 15-25 µm; renewed internally, rarely sympodial, zoospores discharge saprolegnoid; encysted zoospores 10-12.5 µm in diameter; oogonia abundant, terminal, lateral, intercalary, spherical or pyriform, rarely doliform, 25-120 µm in diameter, mostly 55-65 µm; antheridial branches absent, antheridial cell hypogynous; oospores centric, spherical, 1-20 in number.

It was isolated from Nainital lake during summer, rainy, and spring; from Khurpal and Bhim Tal lakes during spring; and from Naukuchiya Tal lake during winter and spring and when temperature ranged between 15-23.5°C (Khulbe and Bhargava, 1977) and pH between 7.5-8.2 (Khulbe, 1980c & 1981). Misra (1982) also reported it from alkaline waters of Lucknow. Khare (1992) collected it from Dhopeshwar pond, Bareilly.

Saprolegnia irregularis Johnson and Seymour

Svensk Botanisk Tidokrift, Bot., **69** : 89-96; 1975

(Plate 38; Figs. 9-11)

Colony compact, dense; hyphae slender, branched, gemmae abundant, irregular in shape and size, sparingly branched, terminal, intercalary, germinating to form hyphae or converting into oogonia; sporangial mode rarely present, terminal, 288.8-566.6 × 33.3-56 µm, renewed by internal proliferation; zoospores discharge saprolegnioid; encysted spores 8-11 µm; oogonia abundant, variable, terminal, rarely intercalary, 88.8-388.8 × 66.6-112.2 µm, spherical, long, 55.4-77.7 µm in diameter; antheridia not seen; oospores 6-4 in number, variable in shape, but commonly spherical, 16.6-22.2 µm in diameter, centric.

Isolated from soil and alkaline waters (pH 7.5 & 8.5), Chinhat lake, Lucknow (Rai and Misra, 1977b; Misra, 1980 & 1982).

Saprolegnia kauffmaniana Pieters

Bot.. Gaz. 60: 488, 1915

(Plate 36; Figs. 8-10)

Principal hyphae slender, 8-27.5 µm thick at base, sparingly branched; sporangia abundant, cylindrical, 65.5-315 µm long by 26.5-37.5 µm in diameter, renewed by internal proliferation; spores 12.5-13.0 µm in diameter, diplanetic; gemmae few, pyriform or irregular, terminal, sometimes intercalary; oogonia abundant, lateral or terminal, variable in shape from spherical, oval to pyriform, 65-110 µm in diameter, mostly 75-90 µm; oogonial wall smooth, thin; antheridial branches diclinous, simple or branched; antheridial cells tubular or clavate, laterally appressed to the oogonial wall; oospores centric, spherical, 4-20 in number, generally 8-10, 28-33 µm in diameter, predominantly 29.7 µm (Khulbe, 1980).

This species was isolated on maize seeds from Nainital lake on 5 June, 1975 (Khulbe and Bhargava, 1977).

Saprolegnia lapponica Gaumann

Botaniska Notiser, p. 156, 1918

(Plate 38; Figs. 6-8)

Growth delicate, two week old colony 1-1.5 μm on hemp seeds; hyphae slender, straight, sparingly branched, 7.15 μm thick at base; zoosporangia slightly greater in diameter than the hyphae, terminal, repeatedly proliferating from within; ger.imae plentiful, a few, spherical, fusiform or clavate, often in moniliform chains; oogonia generally lateral, rarely terminal borne on a short stalk or sessile, an outgrowth from the oogonial stalk, often proliferating into the oogonium, spherical, oblong, rarely intercalary, usually cylindrical in old sporangium, 22-73 μm in diameter; wall colourless, moderately thick; pits numerous and conspicuous of unequal frequency; antheridia did not develop in the entire culture; oospore numerous, generally filling the entire oogonium, centric, spherical, globose or broadly ellipsoidal, 18-72 μm in diameter (Dayal and Thakurji, 1968b).

It was isolated by Thakurji on 16 Nov, 1966, on hempseeds from pond water behind the B.H.U. Press, Varanasi (Dayal and Thakurji, 1968b) for the first time. Later, it was recovered from Nainital lake (in summer, rainy and spring), Sariya Tal (in summer), Khurpa Tal (in spring), Bhim Tal and Naukuchiya Tal lakes (winter and spring) when temperature ranged between 15-23.5°C (Khulbe and Bhargava, 1977) and pH from 7.4-8.8 (Khulbe, 1980c).

Saprolegnia litoralis Coker

The Saprolegniaceae, p. 54, 1923.

(Plate 38; Figs. 1- 5)

Vegetative growth more vigorous and irregular; hyphae stout, branched reaching up to 67 μm in diameter; sporangia not plentiful, young ones almost cylindrical or more often irregular in diameter; usually curved, repeatedly proliferating; gemmae very abundant, spherical, pyriform, clavate etc., often in chains; oogonia abundant as a rule approaching about 51-145.2 μm in diameter; the large one terminal on main hyphae, others on short lateral branches, usually spherical in shape, furnished with rather a few, very conspicuous pits; oospores spherical or illipsoidal, 3-30, mostly 4-7 in an oogonium, diameter 20.4-49 µm; mostly about 23.8-27.2 µm, often elliptic from pressure; antheridia on every oogonium (one to several), androgynous, monoclinous on short branches, which originate very near the oogonium, frequently, when the oogonium is on a short stalk, arriving from immediately below it (Dayal and Thakurji, 1965).

It was isolated on hemp seeds from waters of Varanasi (Dayal and Thakurji, 1965, 1966; Thakurji, 1967). Dayal and Thakurji (1966) isolated it from some permanent ponds (pH 8-8.5) and cemented ponds (pH 7-7.5) of Varanasi during February to April only (Choudhary and Agarwal, 1980 a).

Khare (1992) isolated it from Dhopeshwar pond, Jalalpur lake and Bareilly club pond, Bareilly.

Saprolegnia mixta var. *asplundii* Gaumann

Botaniska Notiser., p. 155, 1918

(Plate 37; Figs. 1-6)

Principal hyphae slender, 10-50 µm thick at base, moderately branched; sporangia abundant, cylindrical or filiform, 125-488 µm long by 22.4-55.4 µm in diameter, renewed by internal proliferation; spores 10-11.5 µm in diameter, diplanetic; gemmae abundant, variable in shape, single or in chain; oogonia abundant, usually lateral, sometimes terminal, spherical or pyriform, rarely oval, 25-375 µm in diameter, predominantly 95-118 µm; oogonial wall smooth, 1.5-2 µm thick, with numerous conspicuous pits; antheridial branches diclinous or androgynous, small, simple or branched, about half of the oogonia have antheridia; antheridial cells tubular or clavate, mostly simple, sometimes branched, laterally or apically appressed to the oogonial wall; oospores centric, spherical, 1-25 in number, generally 6-10, 15-21 µm in diameter, predominantly 19.5 µm (Khulbe, 1980).

Isolated from Nainital lake (in summer, rainy and spring), Sariyatal (summer), Khurpa Tal (spring), Bhim Tal and Naukuchiya Tal lakes (during winter and spring) when water temperature ranged between 15-23°C (Khulbe and Bhargava, 1977), pH between 7.4-8.2 (Khulbe, 1980 a) and nitrates, sulphates and phosphates in different concentrations (Khulbe and Bhargava, 1983)

Saprolegnia monilifera de Bary

Bot. Zeit., **41** : 56, 1883.

(Plate 38; Fig. 14)

Zoosporangia clavate, renewed cymosely, rarely internally; oogonia spherical or clavate, catenulate or rarely single; oogonial wall smooth, pitted or unpitted; oospores 1-16 in number, centric.

Isolated from fresh water ditches, Varanasi (Thakurji, 1967), and lakes, Gorakhpur (Srivastava, 1985).

Saprolegnia monoica Pringsheim

Jahrb. f. wiss. Bot. *1* : 292, 1858.

The Saprolegniaceae p. 49 & 50, 1923

(Plate 38; Fig. 16)

Main threads straight, dense, primary sporangia clavate cylindrical, spores 9-11 μm in diam.; gemmae plentifully, spherical, fusiform or clavate; oogonia usually borne racemosely, bent or straight; oogonia 45-80 μm in diam., spherical, smooth, with several pits in the membrane; oospores 1-35, mostly 5-10 in an oogonium, centric, 18-30 μm in diam.; antheridial branches monoclinous or androgynous forming antheridia on all oogonia, arising near the same stalk of oogonium to which they are attached; antheridia bent, clavate with concave side applied to the oogonium.

Butler (1911) was the first to isolate it from Dehradun. Later, this species was reported from water samples around I.A.R.I., New Delhi during winter at 10-25°C temperature (Chowdhry and Agarwal, 1980 a & b); Sat Tal, Nainital (Mer *et al.* , 1980); polluted water of Ganga Sagar lake, Jabalpur during December, January, March to September (Mishra *et al.*, 1990). Khare (1992) isolated it from Jabalpur lake, Bareilly Club pond and Badaun road pond, Bareilly.

In Himalayas, this species was isolated on hemp seeds from Khurpa Tal lake on 4 August, 1975 (Khulbe and Bhargava, 1977 & 1980 b), and found to be widely distributed in different habitats situated under various ecological conditions, viz.,

Nainital lake (in summer & rainy seasons), Sariya Tal pond (throughout the year except winter), Khurpa Tal, Bhim Tal and Naukuchiya Tal lakes (in all the seasons except autumn & winter) at temperature 15-20°C (Khulbe and Bhargava, 1977); pH between 7.4-8.2 (Khulbe, 1980 C); Ram Tal (Sat Tal) lake (Mer *et al.*, 1980; Mer and Khulbe 1984); cultivated soils of Nanakmatta (Verma, 1984), Ranikhet, Haldwani, Kilbury, Ramgarh (Forest soils), Almora, Dwarahat, Ganai, Haldwani, Pithoragarh, Ramnagar (Cultivated soils), Bhim Tal, Khurpa Tal, Nainital, Naukuchiya Tal, Sat Tal (Khulbe, 1985); oak forest soils of Naina Peak (Nainital) during rainy, autumn and winter, and sal forest soils of Haldwani in all the seasons at 14-30°C temperature, 2-90 per cent moisture and 5.3-6.9 pH.

Saprolegnia parasitica Coker

The Saprolegniaceae, p. 57-58, 1923

(Plate 38; Fig. 12)

Principal hyphae moderately stout, sometimes slender, 15-118 µm in diam., branched; zoosporangia abundant, cylindrical or irregular, sometimes filamentous, 50-105 µm long by 15-80 µm in diameter, renewed internally and sympodially; encysted zoospores 9-11.5 µm in diameter, mostly 10 µm, diplanetic; gemmae abundant, variable in shape, size and position; mostly cylindrical, terminal, 25-90 × 40-600 µm, produce zoospores or germinate directly after a period of rest; resting gemmae deciduous; sex organs absent.

Srivastava (1985) described sexual structures in this species as follows:

Oogonia usually spares, terminal, lateral or intercalary, clavate, pyriform or irregular, 54-146 × 18-72 µm, pyriform ones mostly 62-75 µm in diameter; oogonial wall unpitted, thin, smooth; oospores subcentric, spherical, filling the oogonium, 2-40 in number, 16-28 µm (average 23-26 µm) in diameter; antheridial branches diclinous, very rarely androgynous, simple, delicate, wrapping about oogonium and the hypha persistent; antheridial cells tubular or clavate, laterally appressed. Dayal (1958) tried to induce sex organs of this species in various media, but remained unsuccessful.

This species was isolated by Chaudhuri and Kochhar (1935)

for the first time from Lahore. Later, several investigators reported it from different ecological conditions.

Under warmer conditions, it was isolated from some ponds of Allahabad (Dayal, 1958), lakes of Gorakhpur (Bhargava *et al.*, 1971), soils and water samples around I.A.R.I., New Delhi and Rampur during winter at 10-25°C temperature (Chowdhry and Aggarwal, 1980 a&b); Hanuman Tal pond, Jabalpur (Agarwal *et al* ., 1989) and Supa Tal lake, Jabalpur (Indira *et al.*, 1988). Khare (1992) isolated it from river, pond and lake of Bareilly.

Subtropical and temperate habitats of Kumaun Himalaya also harboured this species at different places, viz., Nainital lake (in all the seasons except summer), Khurpa Tal (autumn & winter) and Bhim Tal and Naukuchiya Tal lakes (winter) at 10-20°C water temperature (Khulbe and Bhargava, 1977), 7.4-8.4 pH (Khulbe, 1980C) and different concentrations of nitrates, sulphates and phosphates (Khulbe and Bhargava, 1983).

Subsequently, it was collected from water and soil samples of Ram Tal, Sal Tal (in spring, rainy and autumn), Garur Tal (spring to rainy and winter) and Nal Damyanti Tal (in all the seasons except autumn) when temperatures of water and soils ranged from 18-23.5° and 8-25.5°C, respectively.

Further, this species was isolated from cultivated soils of Gaṇai, forest soils of Kilbury, Nainital; cultivated soils of Bhim Tal, Khurpa Tal, Russey; and lake waters of Khurpa Tal, Nainital and Sat Tal (Khulbe, 1985). In an ecological study, this species was found in oak forest soils of Nainital during all the seasons (except winter) at 5-18°C temperature, 30-90% moisture, 1.5-4.5% organic carbon and 5.3-5.8 pH. However, sal forest soils of Haldwani did not show this species (Khulbe, 1991).

For the development of sex organs different isolates of this species were cultured at 5, 10,15, 20 and 25°C temperatures under laboratory conditions, but no structure was observed (Khulbe, 1985).

S. parasitica var. *kochhari* Chaudhuri and Kochhar

Proc. Indian Acad. Sci. II(2) : 137-154

Dense, long hyphae; zoosporangia usually very long and stouter, 50-75 × 400-600 μm long; zoospores usually do not liberate but germinate inside the zoosporangium when growing a fresh.

Parasitic on *Belone, Cancinule* and other fishes. Abundant hyphae covering the dead fish. In lab. *Labio rohita* was infected by means of zoospores, but *Ophiocephalous punctatus* injured at the mouth from which hyphae grew out, recovered and lived for 3 months.

Microtome sections of the infected fish showed the presence of the fungus throughout the body of the host.

Collected by S.R. Kashyap from Sheikhupura, Punjab (Chaudhuri and Kochhar, 1935).

Saprolegnia rhaetica Maurizio

Proc. Indian Acad. Sci. *15(4)* : Sec. B, 1942

(Plate 39; Figs. 1-7)

Hyphae branched, 30.4 μm thick, zoosporangia branched, and some branches come out of the empty primary zoosporangia; zoospores 8-9 μm in diam.; gemmae complicated; oogonia 72 × 61.6 μm containing 9-12 eggs, generally 12-19.2 μm in diameter; oogonial wall not very thick, or pitted, pits very few (usually 2-3); antheridia absent.

It was collected on house fly from Lahore in February, 1939. (Chaudhuri and Banerjee, 1942).

Saprolegnia subterranea (Dissman) Seymour

Nova Hedwigia **19** : 59-62, 1970.

(Plate 40; Figs. 1-9)

Hyphae slender, aseptate, branched, 7.5-40 μm in diameter, predominantly 12.5-18 μm; primary zoosporangia abundant, cylindrical, rarely clavate, often slightly curved, 165-630 μm long, mostly 300-450 μm and 7.5-45 μm in diameter, predominantly 25-35 μm, secondary zoosporangia sympodial or in basipetal succession, occassionally by internal proliferation; zoospores liberation saprolegnoid; encysted primary zoospores 7.4-9 μm in diameter; gemmae abundant, variable in shape and size, act as zoosporangia; oogonia abundant, lateral, spherical or pyriform, sometimes within the empty zoosporangium; oogonial wall variable from 10-25% papillate to 50-80% smooth-walled, 24.5-70 μm in diameter, predominantly 30.0-39.6 μm; antheridia mostly androgynous, rarely hypogynous; oospores spherical, centric or subcentric, very rarely centric, single in 96% oogonia and two in 4% oogonia, 21.0-34.8 μm in diameter, mostly 26-30 μm (Khulbe 1985).

This species was isolated from Ram Tal (Sat Tal) soils on August 20, 1979 for the first time (Mer and Khulbe, 1984). The isolate was characterized by slender hyphae; secondary sporangia in basipetal succession, 40-50 per cent papillate oogonia, and presence of hypogynous antheridia in addition to androgynous antheridia. Later, this species was reported from soil samples of Nainital forest (Khulbe, 1985) and found to be distributed in Naina Peak and Kilbury Oak forest of Nainital during all the seasons (temperature 5°C, moisture 30-90%, pH 5.3-6.9, organic carbon 0.8-4.9% (Khulbe, 1991). It was completely absent in sal forest soils of Haldwani (Khulbe, 1991). Khare (1992) isolated this species from Nakatiya river, Jalalpur lake, Bareilly Club pond and Badaun road pond of Bareilly.

Saprolegnia terrestris Cookson ex Seymour.

Proc. Roy. Soc. Victoria (N.S.), 49(2) : 235, 1937

Nova Hedwigia Z. Kryptogamenkd. 19 : 1-124, 1970.

(Plate 38; Fig. 21)

Hyphae slender, 8-40 μm in diameter, branched; zoosporangia abundant, filiform or clavate, 50-350 × 15-50 μm; renewed internally or cymosely; zoospore discharge saprolegnoid; encysted spores 5-10 μm in diameter; gemmae less frequent, variable in shape and size, function as zoosporangia; oogonia abundant, lateral, terminal, intercalary, spherical or pyriform, 30-500 μm in diameter, mostly 50-60 μm; oogonial wall pitted or unpitted, smooth, rarely with a few papillae; antheridial branches androgynous, develop just near the oogonium, rarely branched, antheridial cells simple, clavate; oospores 1-15 in number, mostly 3-6, 18-45 μm in diameter, mostly 25-30 μm, subcentric, very rarely centric.

This species was recorded from soil and water samples of Rampur for the first time (Chowdhry and Agarwal, 1980 b).

Saprolegnia uliginosa Johannes

Arch. Mikrobiol., *14* : 599, 1950

(Plate 38; Fig. 17)

Vegetative growth moderately branched, hyphae thin and flaccid, 16-50 μm thick; gemmae rare; zoosporangia spherical, cylindrical, clavate; 188-344 × 33-67 μm, renewed intercalary; encysted spores 5.5-11 μm in diameter, some spores sprout inside the sporangium, variable; oogonia abundant, variable, lateral with a basal cell, spherical with short stalk, 36-66 μm in diameter, smooth, unpitted; antheridia monoclinous or diclinous; antheridial cells laterally appressed; oospores 2-8 in number, spherical, 14-36 μm in diameter, centric and subcentric.

This species was isolated on hemp and mustard seeds from submerged twigs in central pond, Lucknow University for the first time (Rai and Misra, 1977b; Misra, 1980, 1982). Later, it was reported from fresh water (Temp. 26°C, pH 7.4-8.0) lake of Engineering College, Jabalpur, during February, 1978 and 1979 (Specimen HCIO No. 33745) (Hasija and Khan, 1983); Robertson lake, Jabalpur (Agarwal *et al.*, 1989); from decaying plant material of Ganga Nagar, Jabalpur in all the months except October, January and September (Mishra *et al.*, 1990).

Saprolegnia unispora (Coker and Couch) Seymour

Nova Hedwigia, Z. Krypto. **19** : 57, 1970

(Plate 38; Fig. 13)

Hyphae stout, irregular, sparingly branched, 10-30 μm in diameter; gemmae variable, functioning as zoosporangia; zoosporangia abundant, fusiform, filiform or pyriform, variable in length and diameter, renewed sympodially; zoospores discharge saprolegnoid; rarely dictyoid; encysted spores 10-12 μm in diameter; oogonia abundant, lateral, sometimes terminal or intercalary, frequently clustered on a hypha, spherical or pyriform, 18-68 μm in diameter; oogonial wall pitted or unpitted, smooth, very rarely shortly papillate; oogonial stalks variable in length, rarely branched; oospores subcentric or centric, spherical, 1-4 in number, 18-43 μm, average 32-38 μm in diameter; antheridia absent.

This species was isolated frcm fresh water pond of Allahabad (Dayal, 1958), fresh water and damp soil of Gorakhpur (Srivastava, 1985), from decaying plant materials of polluted water of Ganga Sagar, Jabalpur in all the seasons except October, January and September (Mishra *et al.*, 1990) and from Brahma Sarovar, Kurukshetra (Gupta and Mehrotra, 1988a, 1989a&b).

Plate-32

Figs. 1-3: *Saprolegnia bhargavii* : 1, zoosporangia, 2, gemmae, 3, oogonium with centric oospores and diclinous antheridia (Khulbe & Verma, 1987).

Figs. 4-6: *S. delica* : 4-zoosporangium with internal proliferation 5, chain of spherical gemmae; 6-oogonium with many oospores and diclinous antheridia (Chowdhry, 1976).

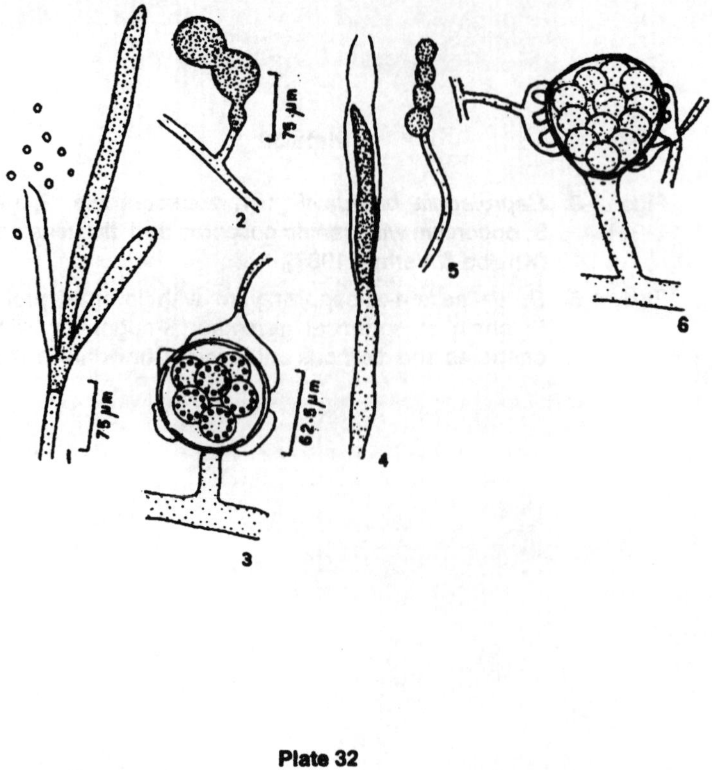

Plate 32

Plate-33

Saprolegnia bhargavii :

Fig .1. Young zoosporangium formed at the tip of hypha.

2. Lateral branch of secondary zoosporangium.
3. Lateral secondary zoosporangia in clusters of 3 or 4.
4. Spherical oogonia like gemmae and one oogonium.
(Photographs by Dr. B. L. Verma).

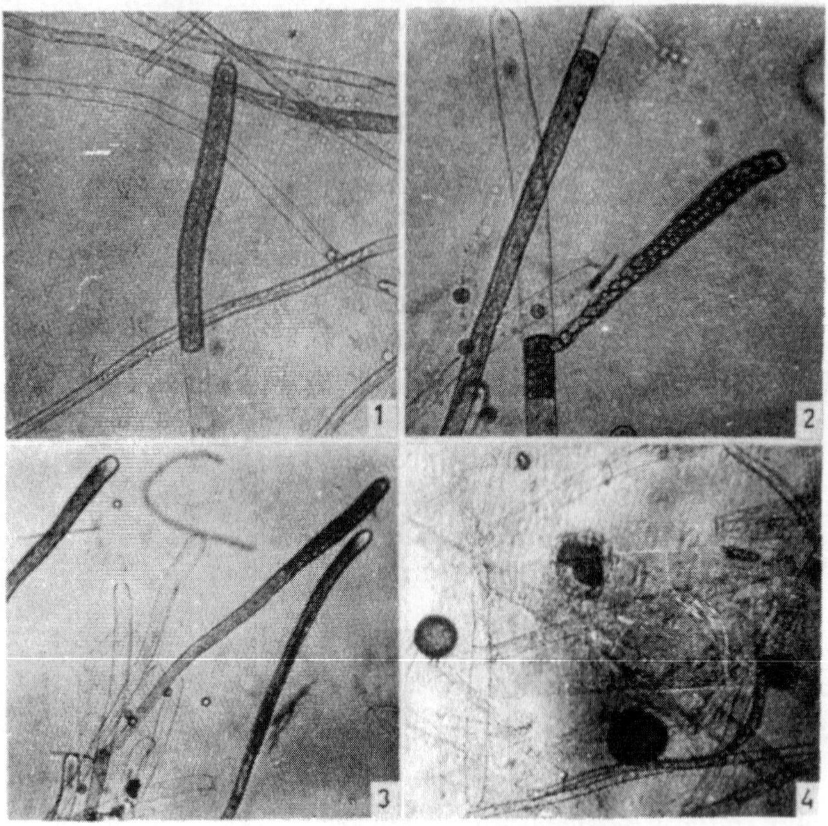

Plate 33

Plate-34

Saprolegnia bhargavii :

Fig. 5- Gemmae in chain.

6. Young oogonium with monoclinous antheridium.
7. Mature oogonium with monoclinous and diclinous antheridia.
8. Oogonia with diclinous antheridia.

(Photographs by Dr. B. L. Verma).

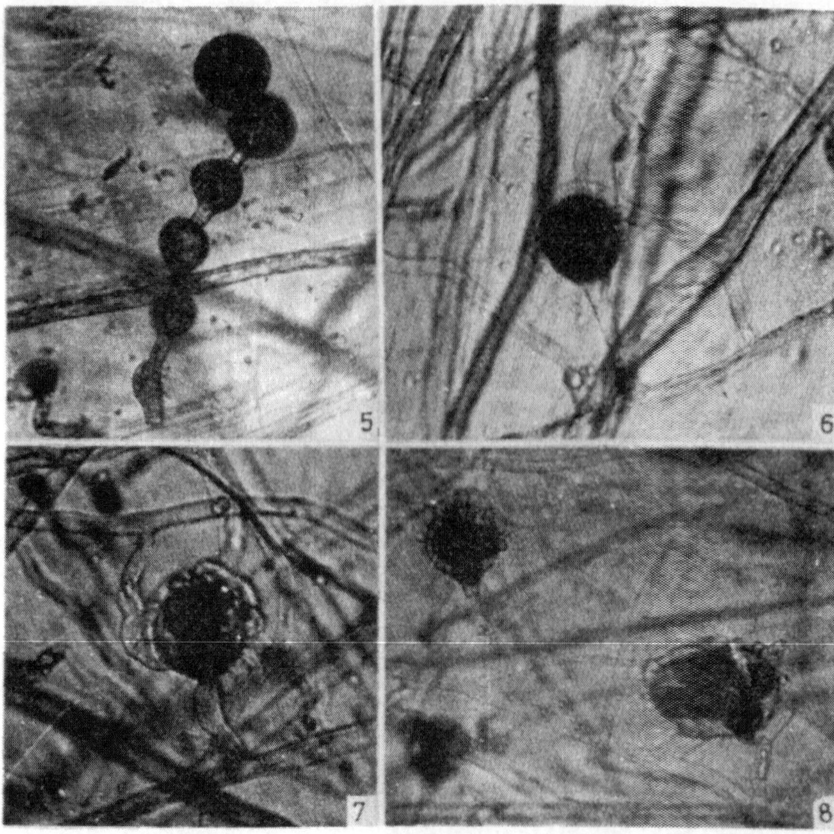

Plate 34

Plate-35

Figs. 1-3: *Saprolegnia diclina* Humphrey.

1. An empty zoosporangium showing internal proliferation.
2. Zoosporangium showing liberation of zoospore.
3. Mature oogonium with centric oospores and a diclinous branch of antheridium.

Figs. 1&2, scale A; Fig. 3, scale B.

Figs. 4-7: *Saprolegnia ferax* (Gruith.) Thuret.

4. An empty zoosporangium showing proliferation into oogonium with oospores.
5. Mature oogonium with centric oospores and monoclinous branch of antheridium.

6&7. Different types of oogonia.

All Figs., scale A.

Figs. 8&9: *Saprolegnia glomerata* (Ties.) Lund.

8. Gemma.
9. Mature oogonium with centric oospores and androgynous antheridial branch.

Fig. 8, scale A; Fig. 9. scale C.

Figs. 10-15 : *Saprolegnia parasitica* Coker.

10&11. Zoosporangia showing internal proliferation

12-15. Gemmae.

All Figs., scale A.

(Khulbe, 1985)

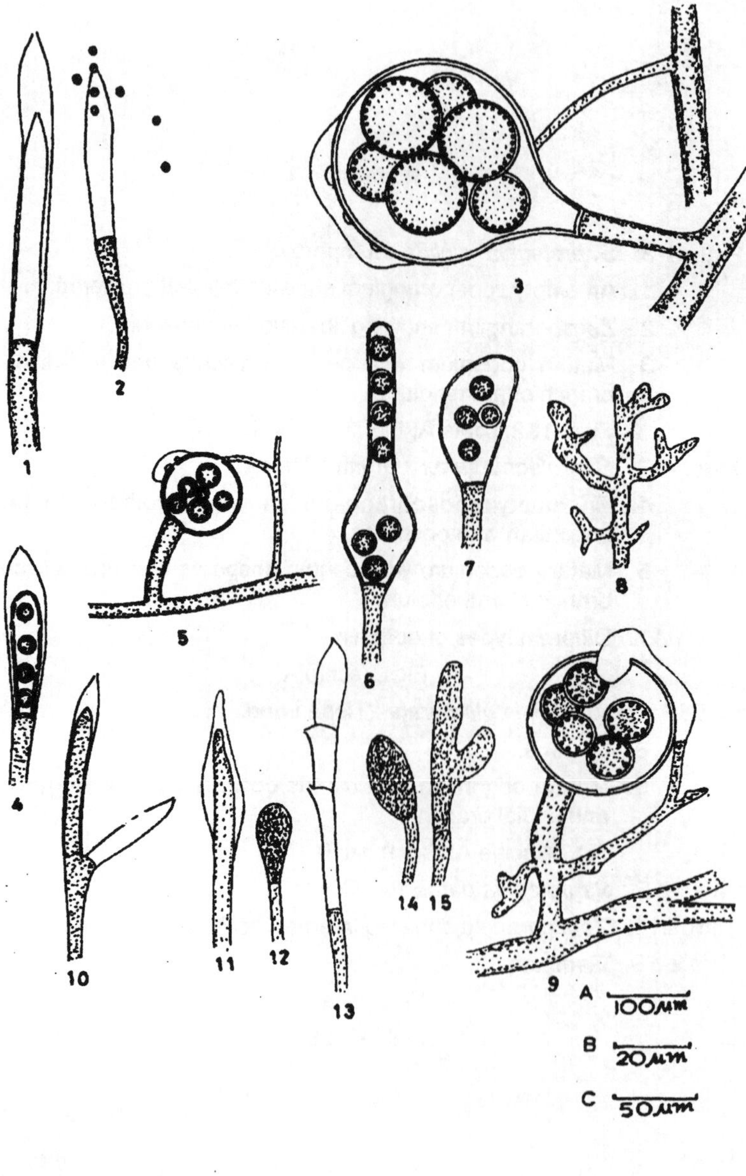

Plate 35

Plate-36

Figs. 1&2 *Brevilegnia linearis* : 1, vegetative filament with one young and other mature sporangium; 2, oogonium with basal antheridium. Figs. 1&2, scale A.

Figs. 3-7 *Saprolegnia crustosa var. III* : 3, proliferating sporangium; 4, gemma; 5, lateral oogonium with diclinous antheridia; 6, terminal and cylindrical oogonium; 7, mature oogonium with centric oospores. Figs. 3-6, scale B; Fig. 7, scale C.

Figs. 8-10 *Saprolegnia kauffmaniana* : 8, proliferating sporangium; 9, terminal oogonium with diclinous antheridia; 10, mature oogonium with centric oospores. Figs. 8&9, scale B; Fig. 10, scale A.

(Khulbe, 1977)

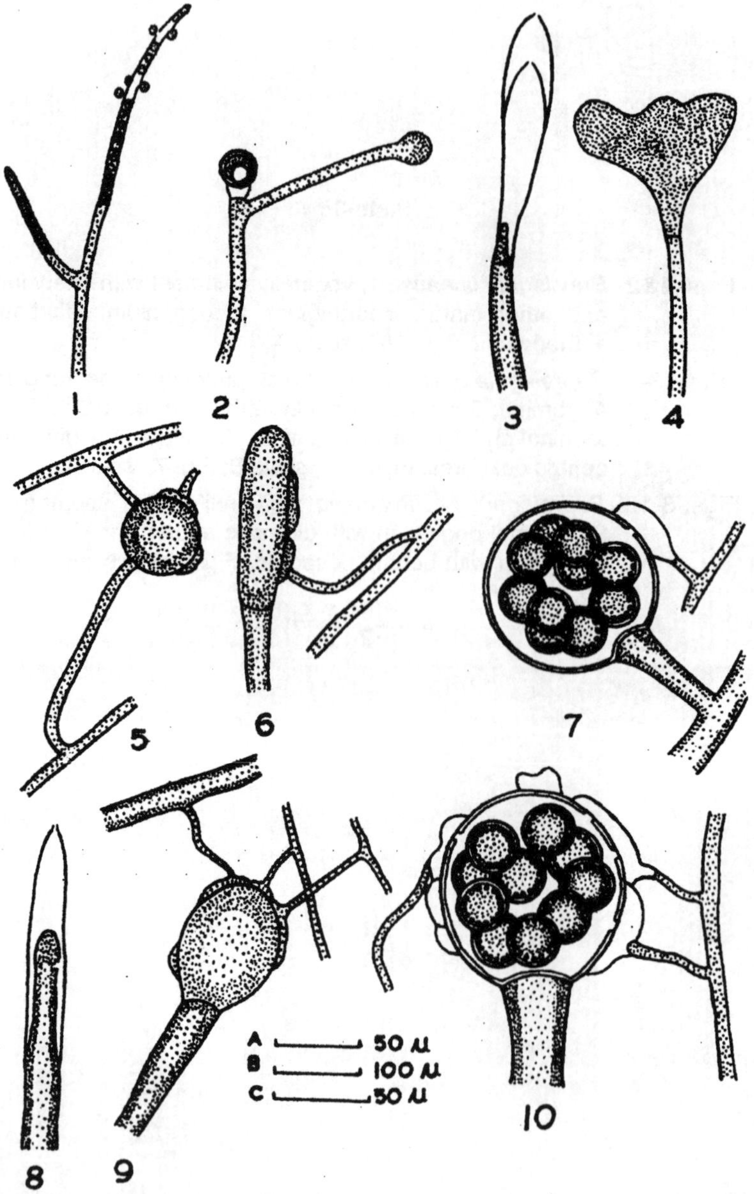

Plate 36

Plate-37

Figs. 1-6 *Saprolegnia mixta var. asplundii* : 1&2 proliferating sporangia; 3&4, gemmae; 5, lateral oogonia with diclinous and androgynous antheridia; 6, mature oogonium showing centric oospores. Figs. 1-5, scale A; Fig. 6, scale B.

(Khulbe, 1977)

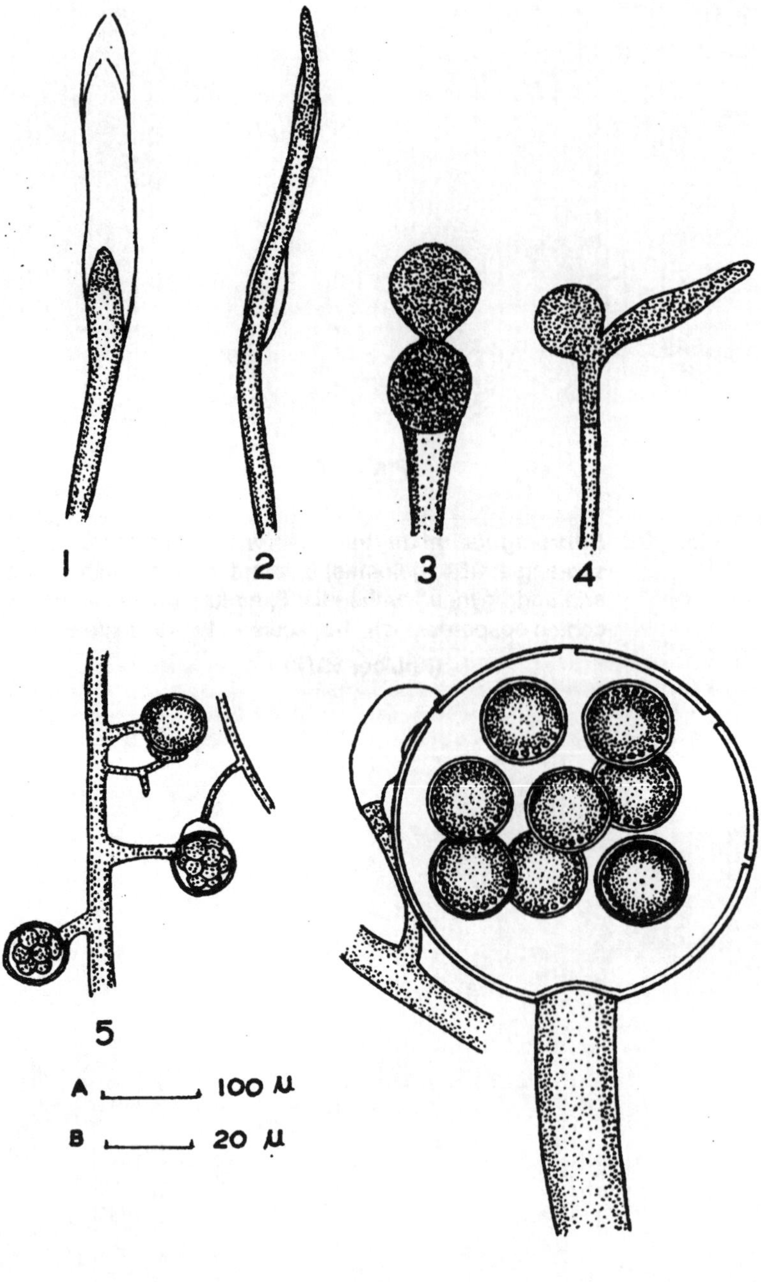

Plate 37

Plate-38

Figs. 1-5: *Saprolegnia litoralis* -1, zoosporangia with internal proliferation; 2-4- gemmae; 5-lateral oogonium with androgynous antheridium (Dayal & Thakurji, 1965).

Figs. 6-8: *S. lapponica* -6, gemmae; 7 & 8- oogonia without antheridia, a large number of oospores, many conspicuous pits and no antheridium (Dayal & Thakurji, 1968b).

Figs.9-11: *S. irregularis* -9, filamentous oogonium with many oospores and no antheridium; 10 & 11-irregular oogonia with many oospores (Misra, 1980).

Fig. 12: *S. parasitica* -oogonium with subcentric oospores and antheridium (Srivastava, 1965).

Fig. 13: *S. unispora* -oogonium with 2 centric oospores (Srivastava, 1965).

Fig. 14: *S. monilifera* -Chain of oogonia with many centric oospores, few pit and no antheridia (Khulbe, 1977).

Fig. 15: *S. anisospora* - Lateral oogonium with variable, eccentric oospores and diclinous antheridia (X 188).

Fig. 16: *S. monoica* -Oogonium with small stalk and monoclinous antheridium (X 247).

Fig. 17: *S. uliginosa* - Lateral oogonium with hypogynous and monoclinous antheridia, centric oospores (Misra, 1980).

Figs.18&19: *S. eccentrica* -Many lateral stalked oogonia with single, eccentric oospores and no antheridia.

Fig. 20: *S. hypogyna* -Lateral oogonium with centric oospores and hypogynous antheridium.

Fig. 21: *S. terrestris* - Oogonium with outgrowth, centric oospores and androgynous antheridium.

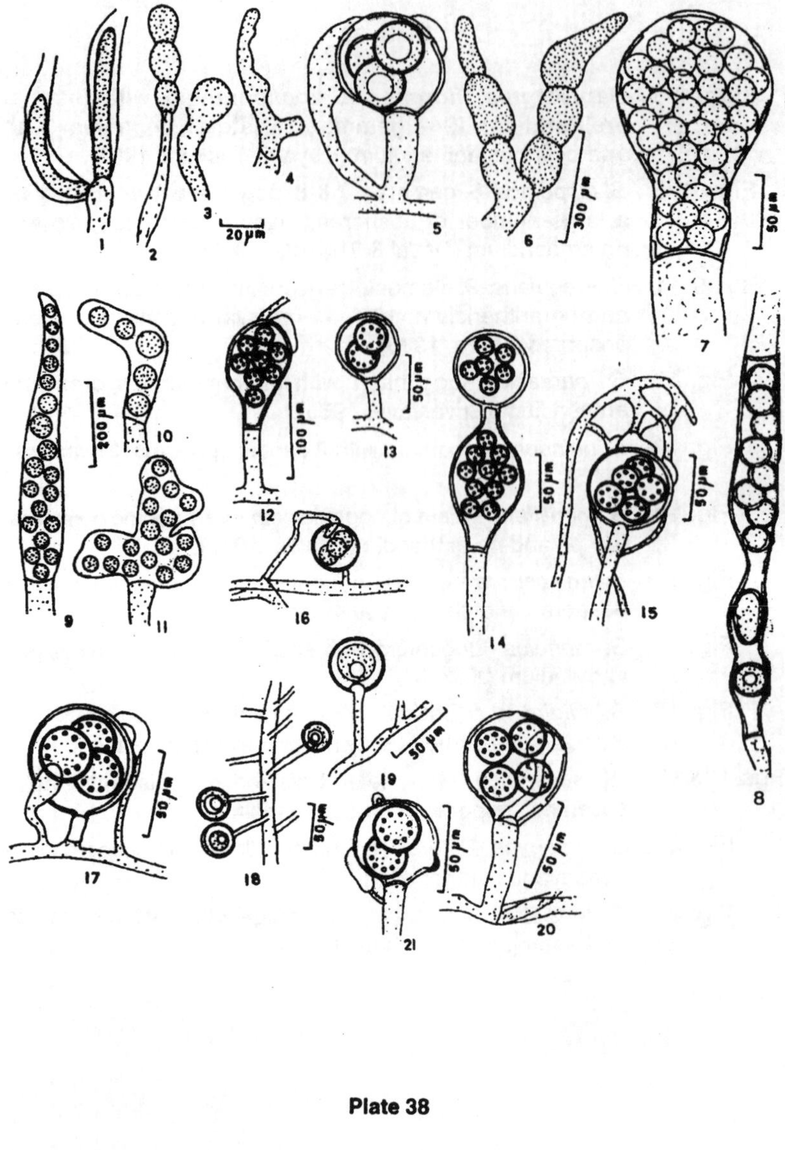
1
2
3
20 μm
4
5
300 μm
6
50 μm
7
8
9
200 μm
10
11
100 μm
12
50 μm
13
50 μm
14
50 μm
15
16
50 μm
17
50 μm
18
50 μm
19
50 μm
20
50 μm
21

Plate 38

Plate-39

Saprolegnia rhaetica

Fig. 1. Internal proliferation of zoosporangium X 105.

2. Secondary zoosporangium X 105.
3. A portion of mature zoosporangium X 487.
4. Complicated gemmae X 105.

5-6. Oogonia with 12 eggs X 487.

7. Oogonium with 9 eggs X 487.

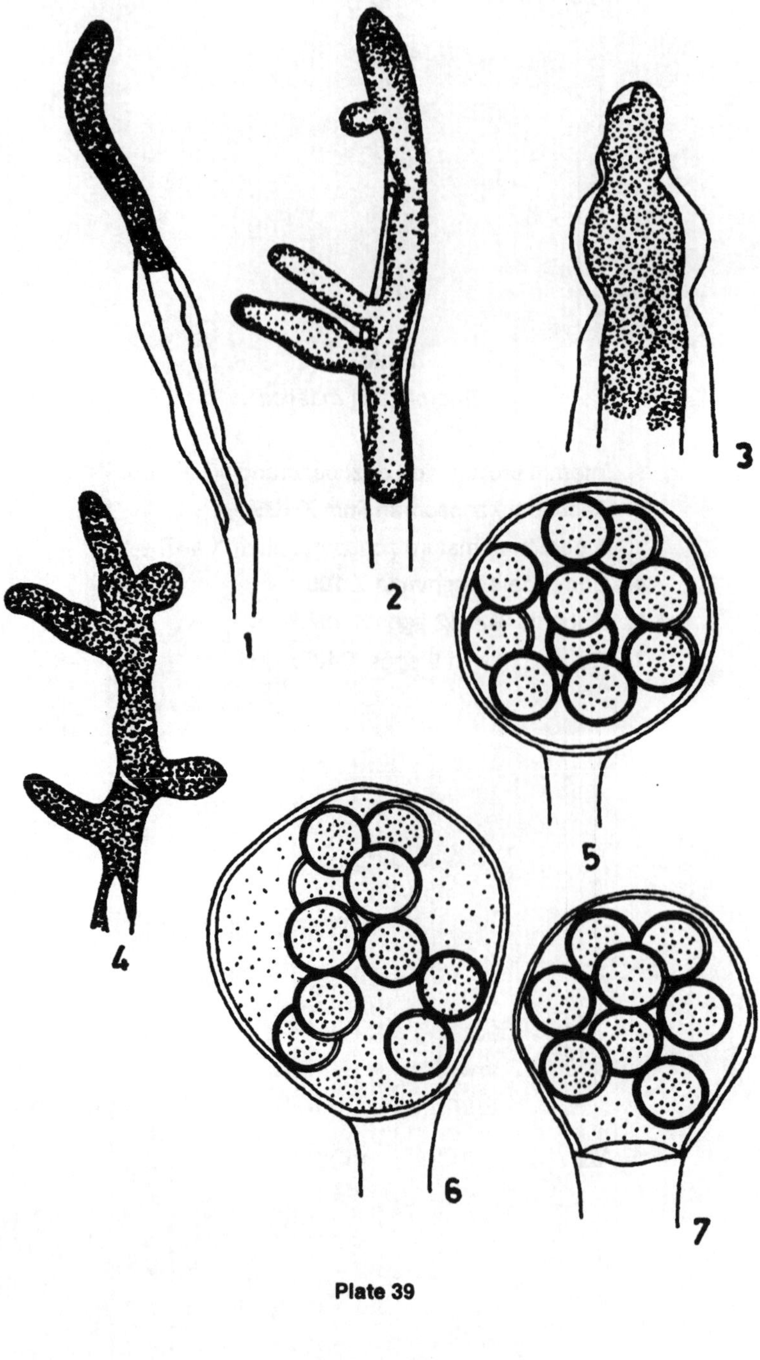

Plate 39

Plate-40

Figs. 1-9: *Saprolegnia subterranea* (Dissmann) Seymour.

1-3. Hyphae showing different types of gemmae.

4. Empty zoosporangium.
5. Zoosporangium showing saprolegnoid liberation of zoospores.
6. Hypha showing lateral stalked oogonium with single centric oospores and absence of antheridium.
7. Hypha with long stalked oogonium and androgynous antheridium.
8. Hypha showing oogonia of different period.
9. Hyphae showing irregular surface, and oogonium with androgynous antheridium.

All Figs., same scale.

(Khulbe, 1985)

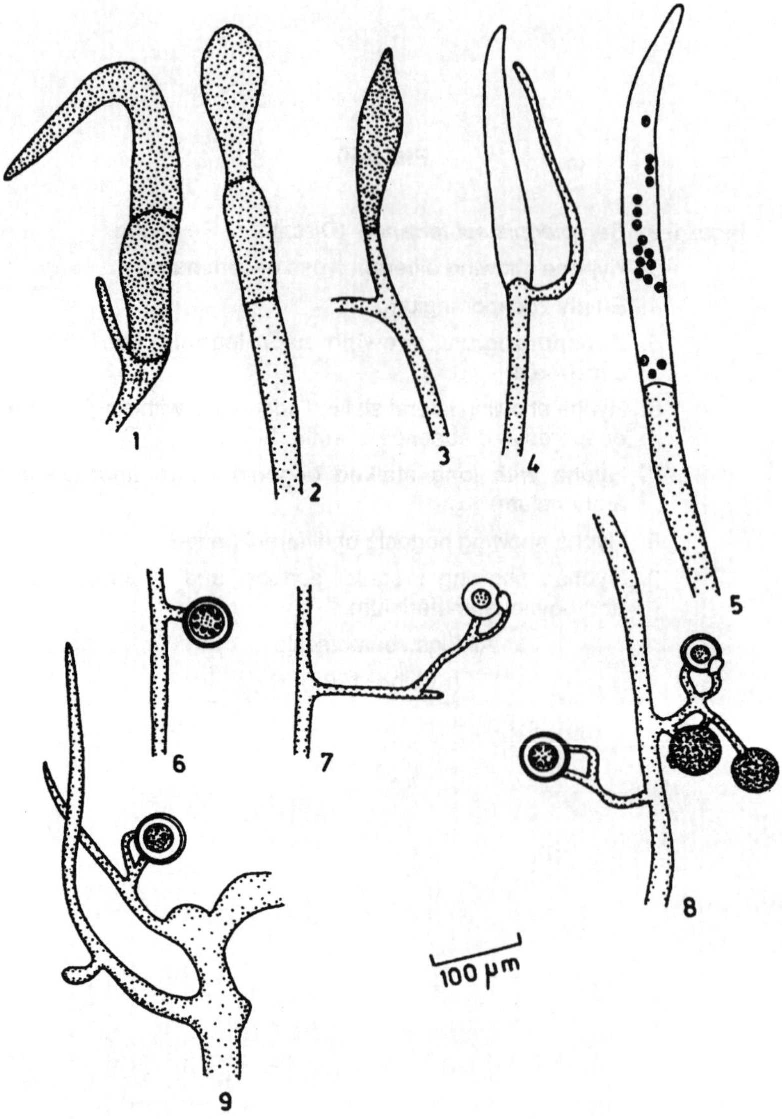

Plate 40

Scoliolegnia Dick

Bot. J. Linn. Soc., **62** : 255-266, 1969

Scoliolegnia centrica Khulbe

Sydowia, Annales Mycologici Ser. II. **36** : 150-153, 1983

(Plate 41; Figs. 1-14)

Vegetative hyphae delicate, flaccid, flexous and branched; zoosporangia elongate, abundant even in old culture, occassionally curved, cylindrical, long, clavate; secondary sporangia usually formed by cymose branching below older one, rarely internal proliferation; zoospores diplanetic, leaving zoosporangium through small pore or long papilla; gemmae abundant, various in shape, behave as zoosporangia; oogonia abundant, borne on slender stalk of variable length, lateral, very rarely intercalary, 30-60 μm in diam., wall irregularly papillate; papillae many, 50-80 per cent, mostly 2-9 in number, 5-28 μm long and 5-7 μm thick, pits absent, present only at the antheridial contact; antheridia present on 50-80 per cent oogonia, androgynous or semihypogynous, antheridial branch arising singly; oospore single, very rarely two, aplerotic, 20-30 μm in diam., mostly 25 μm, centric (Khulbe, 1983).

This species was isolated for the first time from Oak forest of Naina Peak and Kilbury near Nainital on August 18, 1980. The habitats were characterized by low temperature (–5°C–10°C), high water content (45%), low pH (6.5-7.5) and very high humus content.

The present isolate was characterized by abundant gemmae, very often androgynous antheridia, lateral oogonia with few to many irregular and variable papillae and centric oospore (Khulbe, 1983c & 1985).

The type culture was deposited in RDK culture collection, Department of Botany, Nainital (Sc/1) and isotype at CMI, England (IMI No. 262824).

Sommerstorffia Arnaudow

Flora, **116** : 109; 1923

Mycelium fine, not well developed, parasitic in the bodies of rotifers and extending out into the water; some branches with

Plate-41

Figs. 1-14: *Scoliolegnia centrica* Khulbe

1. Single hyphal tip.
2. Young zoosporangium.

3&4. Mature zoosporangia showing liberating zoospores and development of secondary zoosporangia.

5-10. Gemmae.

11&12. Mature oogonia with single centric oospores and androgynous antheridial branches.

13&14. Mature papillate oogonia with single centric oospore and monoclinous antheridial branches.

Figs. 1-12, scale A; Figs. 13 & 14. scale B

(Khulbe. 1985)

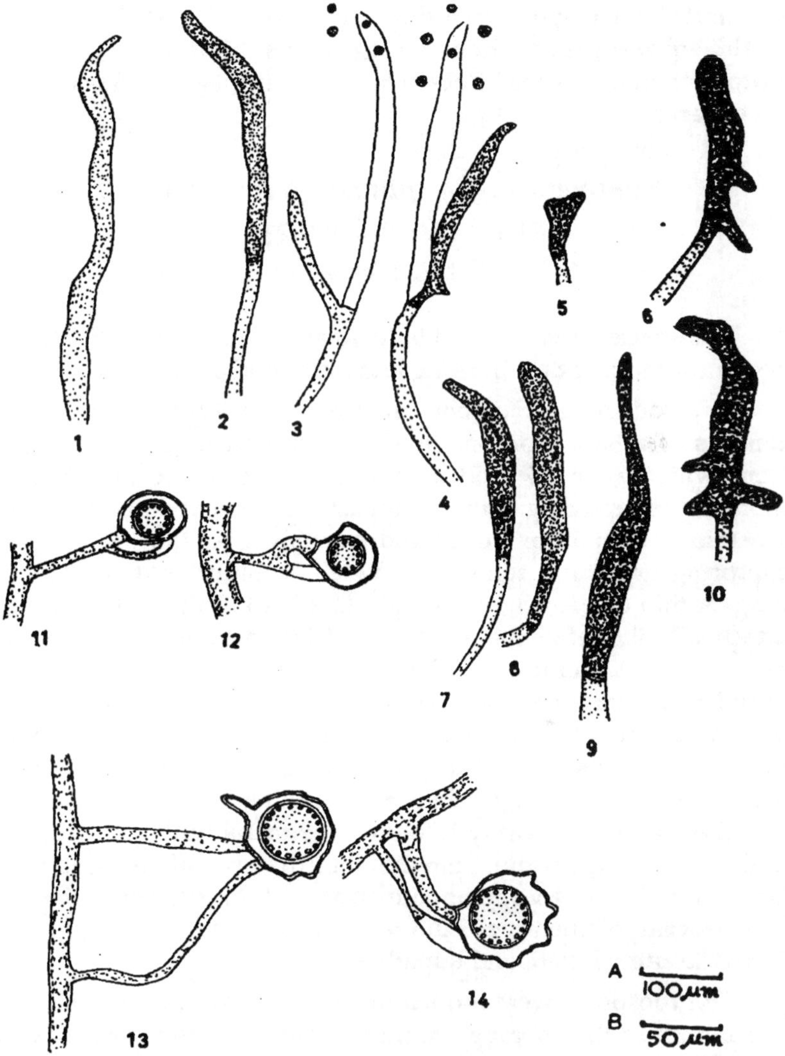

Plate 41

sharp tips covered with mucilage for capturing rotifers; sporangia long and slender; spores in a single row, encysting at the mouth of the sporangium, then emerging as biciliate spores; reproduction by means of oogonia containing egg; antheridium not observed.

Sommerstorffia spinosa Arnaudow

Flora **116**: 109-113, 1923

(Plate 42; Figs. 1-3)

This species was reported by Patil and Kapadnis (1977) from Poona for the first time in India, and described as follows:

Parasitic in the members of rotifers, A typical specimen consists of a basal portion, which is attached to the baits, and from which arise 3-7 isodiametric branches 5-7 × 58-117 µm, modified as capturing organs; sometimes a detached plant is seen wherein it is more branched and extensive. At the tip of each capturing organ is a refractive pointed or blunt part, 10-22 µm long. A thin crozier like or straight (1 × 7-18 µm)mucilage tip is attached to the refractive pointed part. These extension were not observed by Arnaudow (1923) and Sparrow (1929), but they were found to be quite common by Patil and Kapadnis (1977). They also mentioned that mucilage tip might be present on one and lacking on another peg of the same thallus. These mucilage tips were often obscured by debris or by the captured rotifers, but could be seen quite easily in clean specimens without staining, and it was by those mucilage tips that the animals appeared to be captured. The zoosporangium was endobiotic, consisting of isodiametric filaments, 5-6 µm wide, projecting 90-100 µm out from the animal (Patil and Kapadnis, 1977).

The zoospores were formed in a single row, and encysted at the mouth of the sporangium immediately on emergence, each cyst measuring 5-6 µm in diameter. Sexual reproduction was not observed.

It was parasitic in the rotifers in Ram River at Pashan, near Poona, Maharashtra State on 28th June, 1976 (Patil and Kapadnis, 1977).

Plate-42

Figs. 1-3: *Sommerstorffia spinosa* :

1. Thallus inside the rotifer (*Monostyle* sp.)
2. Empty sporangium, one discharge tube with attached primary zoospores cysts.
3. Predaceous peg with hyaline matarix at its tips.

(Patil and Kapadnis, 1977).

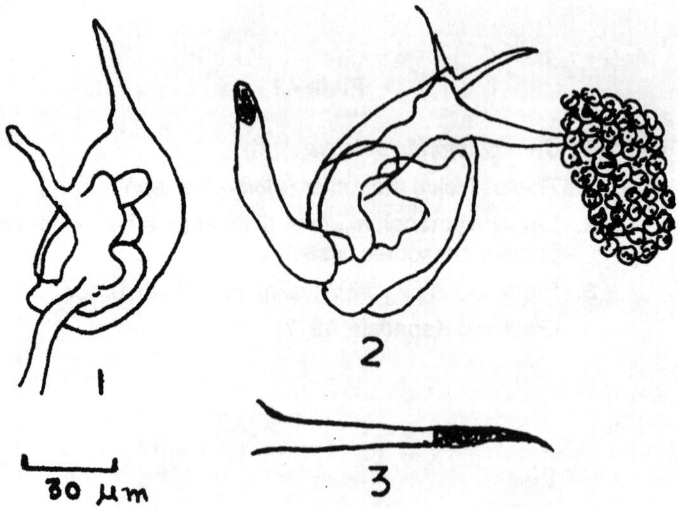

Plate 42

Thraustotheca Humphrey

Trans. Amer. Phil. Soc. *17* : 131, 1892 (1893)

The Saprolegniaceae, p. 148, 1923

***Thraustotheca clavata* (de Bary) Humphrey**

Trans Amer. Phil. Soc., **17** : 131, 1892, (1893)

(Plate 43; Figs. 1-28)

Principal hyphae stout, 10-80 μm thick at base, moderately branched; zoosporangia abundant, clavate, sometimes subcylindrical, spherical, filamentous, 55-500 μm long by 30-100 μm in diameter, renewed sympodially, mostly 150-200 x 30-50 μm; zoospores spherical, 9-12 μm in diameter, mostly 11.5 μm, diplanetic, liberate slowly by disintegration of zoosporangial wall, mostly from middle portion, rarely from basal or apical region; gemmae few, terminal, spherical, clavate, pyriform; oogonia abundant, spherical to ovoid, 40-150 μm in diameter, mostly 48-60 μm, lateral on short to long stalks; oogonial wall 1.5-2 μm thick with many conspicuous pits; antheridial branches diclinous, simple or branched, 1-3 antheridia per oogonium; antheridial cells club-shaped, laterally appressed or attached by few to many projections to the oogonial wall; oospores spherical, eccentric, 1-10 in number, mostly 6-8, 15-50 μm in diameter, predominantly 23 μm (Khulbe, 1985).

Isolated on hemp seed from cultivated soils of Dwarahat on March 7, 1980, leg. R.D. Khulbe. The type culture has been deposited in the Department of Botany, Kumaun University, Nainital (T/1).

This species was isolated from soils of Dakha, Ludhiana for the first time (Chaudhuri and Lotus, 1936). Later, it was reported from Bhimtal and Naukuchiya tal lakes, Nainital (Khulbe, 1980a & 1985); soil samples form Government House, Ayarpata, Tiffin Top, Nainital (Sati and Khulbe, 1980); non rhizosphere soils of pea around Nainital (Khulbe, 1983); Ram Tal and Sat Tal waters (Mer and Khulbe, 1984); cultivated soils of Dwarahat, forest soils of Kilbury and Naina Peak (Khulbe, 1985).

This species was found to be flourished under various ecological conditions in Kumaun Hills, viz., temperature range, 15-23.5°C (Khulbe and Bhargava, 1977), 19-25.5°C (Mer *et al.*, 1980),

5-18°C (Khulbe, 1985); hydrogen-ion concentration from 7.4-7.5 (Khulbe, 1977), 5.3-6.8 (Khulbe, 1985), moisture contents between 30-90% and organic carbon from 0.8-4.9% (Khulbe, 1985), during autumn (Khulbe and Bhargava, 1977) and in other seasons (Khulbe, 1985).

Plate-43

Figs. 1-28: *Thraustotheca clavata* (de Bary)

1. Single hyphal tip.

2-11. Different types of zoosporangia.

12-16. Different types of gemmae.

17-22. Lateral oogonia on variable stalks.

23. Single oogonium with monoclinous antheridial branch.

24. Oogonium with diclinous antheridial branch.

25. Numerous oogonia with monoclinous antheridial branches.

26. Coiled antheridial branch.

27. Oogonium with diclinous antheridia and developing oospores.

28. Mature oogonium with eccentric oospores and diclinous antheridial branches.

All Figs., same scale.

(Khulbe, 1985)

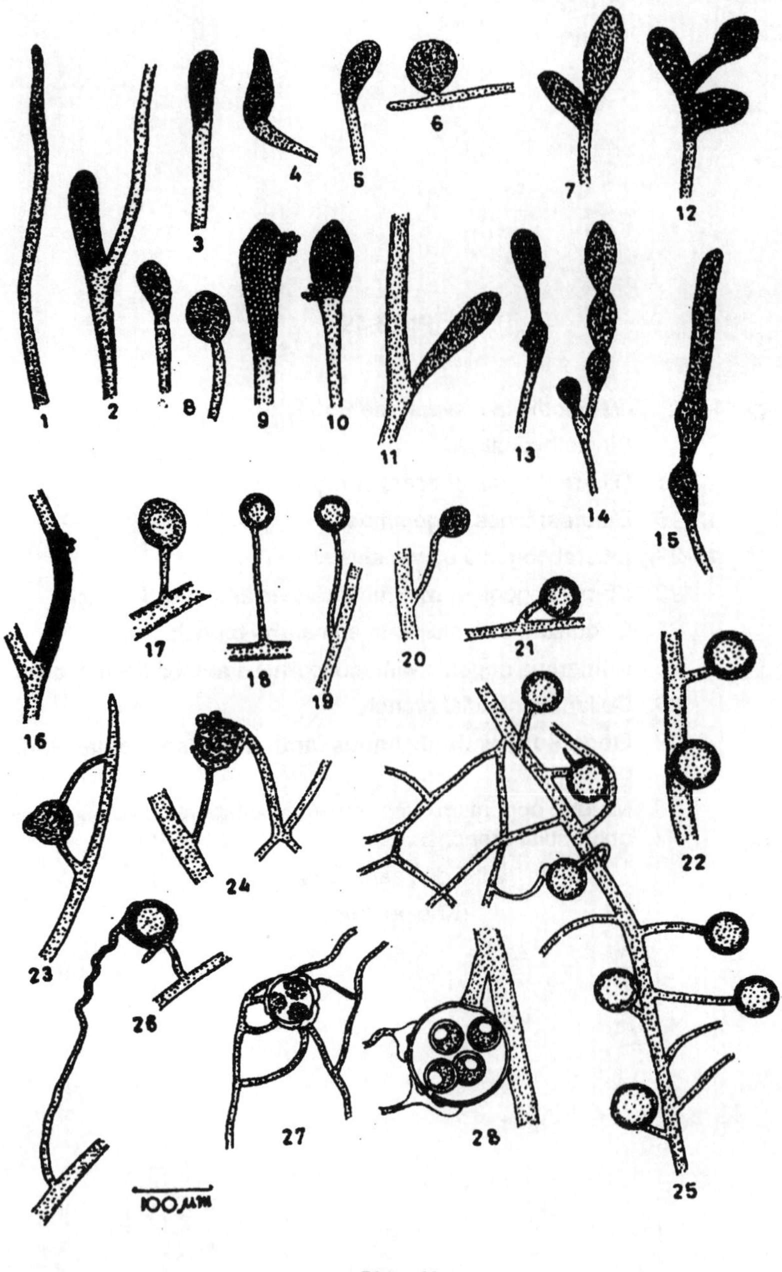
1
2
3
4
5
6
7
8
9
10
11
12
13
14
15
16
17
18
19
20
21
22
23
24
25
26
27
28
100 μm

Plate 43

Plate-44

1. *Achlya klebsiana* : Intercalary and cylindrical gemmae.
2. *Thraustotheca clavata* : Liberation of zoospores.
3. *Brevilegnia subclavata* : Liberation of zoospores.
4. *Geolegnia inflata* : Zoosporangium showing suppression of motility of zoospores.

   (Photographs by Dr. R. D. Khulbe).

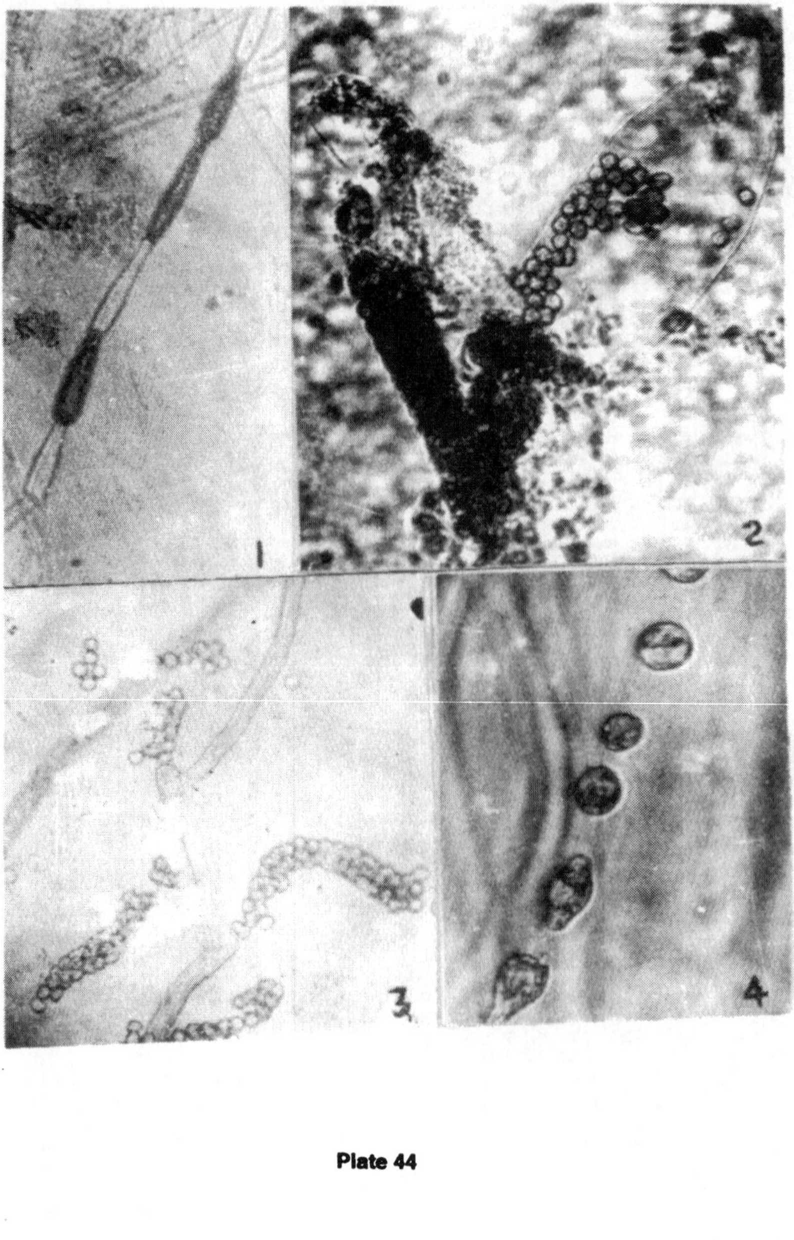

Plate 44

Plate-45

Fig. 1. Zoospores liberation of *Achlya americana.*

2. Zoosporangia & zoospores of *Saprolegnia subterranea.*

3. Mature zoosporangium of *Protoachlya oryzae.*

4. *Brevilegnia diclina* - zoospores liberation

5. *Dictyuchus monosporus* - sporangium

6. *Achlya* sp. of Aplanoid germination of zoospores.

7. *Saprolegnia bhargavii* - showing lateral secondary zoosporangium as *Isoachlya.*

8. *Protoachlya oryzae* - zoosporangium showing many papillae.

9. *Leptolegnia caudata* - single row of zoospores.

10. *Saprolegnia* sp. internal proliferation of zoosporangium.

11-14. *Achlya* spp. various types of gemmae.

15. *Saprolegnia lapponica* - cylindrical oogonium with many oospores and no antheridium.

16. *Achlya americana* - lateral oogonia and monoclinous antheridia

17. *Saprolegnia ferax* - various types of oogonia & many oospores.

18 & 28. *Scoliolegnia centrica* - papillate oogonium with single centric, oospore.

19. *Saprolegnia anisospara* - cylindrical oogonium with various oospores, diclinous antheridia.

20-21. *Dictyuchus pisci* - oogonia, diclinous antheridium, single eccentric oospores.

22. *Dictyuchus anomalus* - oogonia with single oospore & no antheridium (Photo by Srivastava, 1981).

23,24,26,27 - *Achlya* spp. - oogonia with many oospores, various antheridial origin.

25- Oogonium with androgynous and diclinous antheridia of *Dictyuchus pisci.*

(Photographs by Dr. R.D. Khulbe, except photo No. 22).

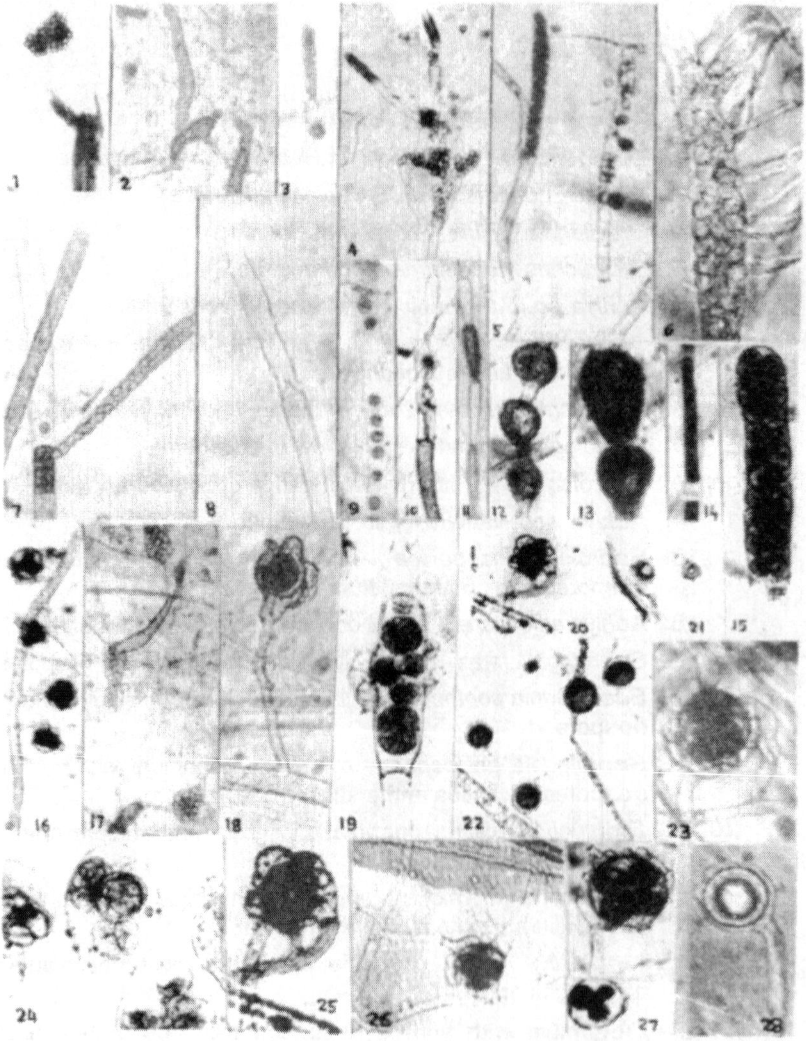

Plate 45

LEPTOMITALES

Leptomitaceae

Apodachlya Pringsheim

The Saprolegniaceae, p. 172, 1923

Apodachlya minima Coker and Leitner var. *hypogyna* (Khulbe) Mer and Sati.

Current Science **56** (17) : 906-907

(Plate 46; Figs. 1-11)

Mycelium flocculent; hyphae slender, constricted into segments of variable length; segments 39.6-181 µm long and 6.6-10 µm thick on hemp seed; branching takes place from any point of segment, mostly from distal part; zoosporangia absent; gemmae absent; oogonia abundant, born on the tips of short moniliform, often recurved lateral branches, spherical, very rarely pyriform, 11.5-20 µm in diameter, predominantly 16.5-20 µm in diameter; oogonial wall smooth and unpitted; sub-oogonial cell spherical, clavate or oval, more dense than other cells of the same chain, acts as an antheridium and becoming empty before the maturation of the egg; oospore single, aplerotic, glutamate with conspicuous oil droplets, eccentric in position as in some eccentric species of *Achlya* (Khulbe, 1985).

Isolated on hemp seed from Sat Tal (Nainital) lake on April 10, 1980, leg. R.D. Khulbe. The type culture has been deposited in the Department of Botany, Kumaun University, Nainital (AP/1) (Khulbe, 1985).

This species was again isolated from the soils of Naina Peak, Nainital (2611 m altitude) (Mer and Sati, 1987).

Plate-46

Figs. 1-11: ***Apodachlya minima*** **var.** ***hypogyna***

1-6. **Vegetative hyphae showing branching pattern and constriction.**

7-8. **Sexual filaments showing terminal oogonium and hypogynous antheridium.**

9-11. **Mature oogonium with hypogynous antheridium; single, plerotic, eccentric oospore with fusion of oil droplets.**

Figs. 1 & 3-6, scale C; Figs. 2, 9 & 10, scale A; Figs. 7 & 8, scale B.

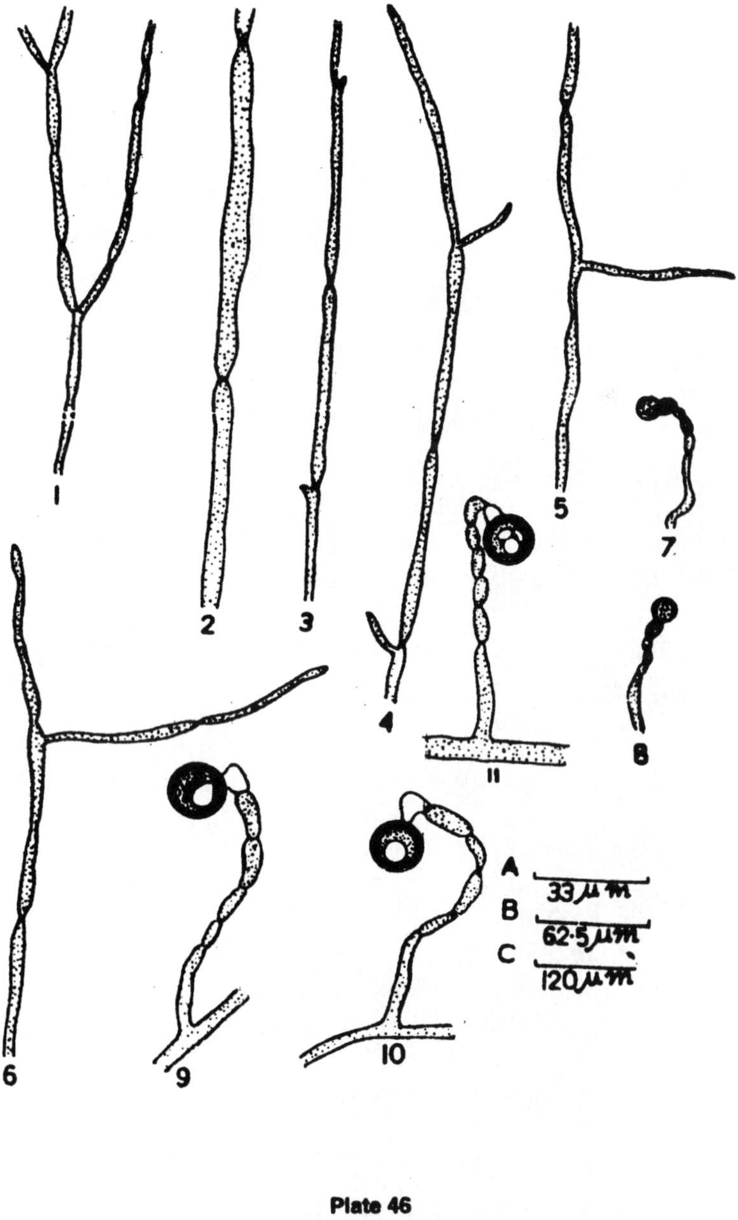

Plate 46

LAGENIDIALES

1. Olpidiopsidaceae
2. Lagenidiaceae

Olpidiopsidaceae

Olpidiopsis Cornu

Ann. Sci. Nat. Bot., V, **15** : 114, 1872

Key to the species of Olpidiopsis

A_1 Companion cell present .... B

A_2 Companion cell absent C

B_1 Resting spores smooth walled .... D

B_2 Resting spores spiny or tuberculated .... E

C_1 Resting spores smooth walled .... F

C_2 Resting spores spiny or warty .... G

D_1 Sporangia and resting spores ranged from 15-135 µm and 30-90 µm, respectively........ .... *O. saprolegniae* var. *levis*, p. 234

D_2 Sporangia 17-27.4 µm, resting spores 3-17 µm...... *O. indica*, p.230

E_1 Exospores tuberculated or wavy .... H

E_2 Exospores spiny I

G_1 Spines delicate, small, 1-1.5 µm long .... *O. pythii*, p. 232

G_2 Small warts present .... *O. aphanomyces*, p. 229

H1 Resting spores mostly ellipsoidal, wall dark grey .... *O. incrassata*, p.232

H_2 Resting spores mostly spherical, wall colourless .... J

I_1 Companion cells smooth walled .... K

I_2 Companion cells thick and with short ornamentations .... *O. varians*, p.235

J_1 Exospores with warts or protuberance........
....*O. saprolegnae var. indica*, p. 233

J_2 Exospores wall show narrow or broad spines......*O. achlyae*, p. 228

K_1 Spines are long, pointed, 5-12 μm L

K_2 Spines small, slender, 1.5 μm .... M

L1 Sporangia mostly spherical, 45-135 μm in diameter, exit tubes 1-3, 15-24 μm in length
....*O. saprolegniae* var. *saprolengiae*, p. 234

L_2 Sporangia mostly ellipsoidal, 60-330 × 30-90 μm, exit tubes 1-3, variable in length .... *O. fusiformis*, p.230

M_1 Spines of resting spores minute, sharp...*O. brevispinosa*, p. 229

M_2 Spines of resting spores conical ....*O. luxurians*, p.231

Olpidiopsis achlyae Mc Larty

Bull. Torrey Bot. Club, **68**: 62, 1941.

(Plate 48; Figs. 6-9 & 18-20)

Zoosporangia solitary or numerous, usually in a terminal swelling of the host filament, spherical, oval, ellipsoidal or elongate, variable in size, ranging from 14-115 × 105-550 μm; 1-3 exit tubes which may extend considerably beyond the surface of the host filament; zoospores numerous, hyaline, small, spherical, oval, 2.0-6.0 × 3-5.5 μm, mostly 3.5 × 3.8 μm; resting spore spherical, 28.8-130 μm; mostly 45 × 40 μm, brown, with one to many refringent globules; companion cells may or may not present (Khulbe, 1985).

Parasitic in *Achlya flagellata* Coker, collected from Nainital lake on March 4, 1983, leg. R.D. Khulbe. The type culture has been deposited in the Department of Botany, Kumaun University, Nainital (0/1). It was also found in Sat Tal lake. Earlier, this species was reported as a parasite of *Achlya* sp. from Lucknow (Das-Gupta and John, 1953), Varanasi (Dayal and Thakurji, 1968b), Kerala (Karling, 1966) and Hyderabad (Manoharachary and Rao, 1981). It was found as a parasite of *Achlya flagellata* , *A. papillosa* and *Saprolegnia* sp. in Pune (Kapadnis, 1980). It could also infect *Thraustotheca clavata* in an artificial inoculation experiment (Khulbe, 1985). Khare (1992) found it parasitic on *Achlya prolifera* and *A. flagellata* in river, pond and lake of Bareilly.

Olpidiopsis aphanomyces (Cornu) Fischer

Rabenhorsts Kryptog. Fl. 1 (4): 37, 1892

(Plate 48; Figs. 1-5)

Sporangia terminal in hyphal swellings, smooth, thin walled, brown in colour, spherical, subspherical or oval, 10-15 µm in diam., 10-18 µm in length; zoospores laterally biflagellate, ovoid, escaping singly through the exit tubes and swimming away immediately, 2-3 × 3 µm in measurements; resting spores spherical, thick walled, warty, 10-12 µm in diam., germination not observed; companion cells lacking.

This was reported as a parasite of *Aphanomyces laevis* and *Aphanomyces* sp. by Karling (1966) from Pusa for the first time.. Later, it was found in *Aphanomyces* sp. in an agricultural pond water, B.H.U., Varanasi by Usha Kiran in September, 1978 (Dayal and Thakurji, 1968b) and *Aphanomyces* sp. from Pune in May, 1977 (Kapadnis, 1980).

Olpidiopsis brevispinosa Whiffen

Amer. J. Bot. **29** : 610.

(Plate 47; Figs. 1-3)

Thallus endobiotic, holocarpic; zoosporangia 1-3 in terminal or intercalary swellings of the host hyphae, 11.5-45 µm in diam., predominantly 21-27 µm, spherical, elliptical; wall smooth, thin, colourless, exit tubes 1-3, variable in length; zoospores numerous, elongated, swim actively; discharge through exit tubes; encysted spores 1.5-2 µm in diam.; resting spores spherical, ovoid, covered with sharp, pointed, small spines, 1-2.5 µm long; resting spores 13.2-39.6 µm in diam.; companion cells 1-2 with each resting spore, spherical, oblong, 16.5-19.8 µm in diam.; wall thin, smooth and colourless (Khulbe, 1980a),

This isolate was found to be parasitic in the hyphae of *Pythium monospermum*, collected from Sariya Tal lake, Nainital on 25th August, 1975 for the first time in India (Khulbe, 1980a).

Olpidiopsis fusiformis Cornu

Ann. Sci. Nat. Bot. V, **15**: 147. 1872

(Plate 48; Figs. 10 & 11)

Thallus endobiotic, holocarpic; zoosporangia one to many in terminal or intercalary swellings of the host hyphae; variable in shape, ranging from spherical, ovoid, fusiform, ellipsoidal to cylindrical, 30-300 µm long by 10-90 µm in diameter; wall smooth, thin, colourless; exit tubes 1-3; variable in length; zoospores many, ovoid, biflagellate; 2-3 µm long by 1.2-1.5 µm thick in diameter; resting spores spherical or ellipsoidal; 40-70 µm in diameter; outer surface covered with many small, pointed, straight or curved, 2–7 µm long spines; companion cells 1-3 with each resting spores, spherical or ovoid, smooth walled, 19-30 µm in diameter.

Parasitic in *Achlya americana* Humphrey, collected from Nainital lake on March 18, 1983, leg. R.D. Khulbe. The type culture has been deposited in the Department of Botany, Kumaun University, Nainital (0/2) (Khulbe, 1985).

Olpidiopsis fusiformis Cornu has been found parasitic in *Achlya imperfecta* Coker (Shanor, 1940), *A. klebsiana* Pieters (Srivastava, 1966) and *A. flagellata* Coker (Khulbe, 1980a). Previously, it was reported as a parasite from Dehradun (Butler and Bisby, 1931); in *Achlya klebsiana, A. polyandra* from Jatasanker Pond, Gorakhpur on 30 October, 1962 (Srivastava and Bhargava, 1963); from Kerala (Karling, 1966); *A. flagellata* in Khurpa Tal lake on March 5, 1974 (Khulbe, 1980a) and from Lucknow (Rai and Misra, 1981). In a host range study this parasite could not infect any other fungal species except *A. flagellata* -the original host (Khulbe and Bhargava, 1981). In the recent past, this species was observed in *A. klebsiana* collected from water of Kurukshetra (Gupta and Mehrotra, 1988b & 1992).

Olpidiopsis indica *Srivastava*

Current Science **44** (17) : 642-644, 1975

(Plate 47; Figs. 6 & 7)

Thallus endobiotic, holocarpic, causing oval, spherical or ballon shaped enlargement at the end of the host hyphae or in lobulate elements of host sporangia, and usually leading to the septation of the hyphae; zoosporangia one to several, mostly oval, spherical or elliptical, variable in size, 17-27.4 µm in diam. (13.6-35 × 17-27.4 µm when elliptical); wall smooth, colourless, discharge tube 1-3, narrowly cylindrical, 3.4-17 µm in length; zoospores oval to elongate, 3.5-5 µm in length, biflagellate, flagella almost equall in length, attached near the apical end; resting spores always with one, two or rarely three companion cells, spherical, 13.6-28 µm in diam., with one or more refringent globules, exospore wall colourless, smooth, varying from 1.7 to 3.4 µm in thickness; companion cells oval or spherical, 6.8-10.5 µm in diam.; wall thin, smooth and colourless (Srivastava, 1975a).

It was found to be parasitic in *Pythium aphanidermatum* (Edson) Fitz. in fallen decaying leaf of *Carica papaya* in Gorakhpur (Srivastava, 1975 a). Type at Botany Department, St. Andrews College, Gorakhpur, S-36 and at I T C C, I.A.R.I., New Delhi.

In a host range study of this obligate parasite, Srivastava (1982) reported that it could parasitize *Pythium ultimum* and *P. carolinianum*.. However, it could not infect *Saprolegnia diclina, S. ferax, Achlya dubia, A. proliferoides, A. orion, A. klebsiana, A. flagellata, Isoachlya anisospora* var. *indica, I. luxuriens, Dictyuchus sterile, D. monosporus, Aphanomyces laevis, A. helicoides, Pythium undulatum* and *Phytophthora gonapodyoides* (Srivastava, 1982).

Olpidiopsis luxurians Barrett

Ann. Bot. London, **26** : 231, 1912

(Plate 48; Figs. 12 & 13)

Thallus endobiotic, holocarpic; zoosporangia 1-12, mostly 2-4 in terminal or intercalary swellings of the host hyphae, spherical, or elliptical, variable in size from 15-80 µm long by 18-100 µm in diameter; wall smooth, thin colourless; discharge tubes 1-5 per zoosporangium, 2-5 µm in length; zoospores numerous, elliptical, biflagellate, small, 1.2-1.5 µm in diameter when encysted; resting spores spherical, 12.5-72.5 µm in diameter, outer surface covered with numerous conical, 8.5-2 µm long spines; companion cells 1-2 with each resting spores, spherical or elliptical, smooth and thick walled, 5-15 µm in diameter (Khulbe, 1985).

Parasitic in *Aphanomyces helicoides* von Minden, collected from Nainital lake on June 15, 1982, leg. R.D. Khulbe. The type culture has been deposited in the Department of Botany, Kumaun University, Nainital (0/3).

Earlier, this species has been reported by Srivastava and Bhargava (1963), Kapadnis (1980) and Khulbe (1980a) as a parasite of *Aphanomyces laevis* , and Srivastava (1966) observed it in *A. helicoides.* In a host range study this parasite could not infect any species of watermolds exept *Aphanomyces laevis* (Khulbe and Bhargava, 1981).

Olpidiopsis pythii (Butler) Karling

Simple Holocarpic Biflagellate Phycomycetes, P. 47, 1942

(Plate 48; Figs. 16 & 17)

Thallus endobiotic, holocarpic; sporangia one to many in terminal or intercalary swellings of the host hyphae; spherical or elliptical; variable in size from 10-39.5 µm in diameter, predominantly 19.8-26.5 µm, wall smooth, thin, colourless; exit tubes 1-3, variable in length; zoospores numerous, oval, biflagellate; encysted spores 1.5-2 µm in diameter; resting spores spherical or oval; 10-36.5 µm in diameter, predominantly 21 µm; outer surface brown, covered with delicate, small, 1-1.5 µm long spines; companion cells absent (Khulbe, 1980a).

Parasitic in the hyphae of *Pythium monospermum*, collected from Naukuchiya Tal lake, 5 March 1972, leg. R.D. Khulbe, (Khulbe, 1980a).

Olpidiopsis incrassata Cornu

Ann. Sci. Nat. Bot., V, 15 : 146, 1872

(Plate 47; Fig. 5)

Thallus endobiotic, holocarpic; zoosporangia one to several, formed in terminal or intercalary swellings of the host hyphae, spherical, subspherical, ellipsoidal or cylindrical, extremly variable in size, 36-155.5 µm long by 14.5-72.6 µm in diam.; wall smooth, colourless, discharge tubes 1-2, variable in length; zoospores oval to elongate, 3.5-5 µm in length, biflagellate; resting spores mostly ellipsoidal, sometimes spherical, 40.5-75 µm long by 30-45.5 µm broad, 45-54.5 µm in diameter when spherical, exospore wall dark grey, warty; companion cells 1-2 with each resting spores, spherical or oval, 18-21 µm in diameter; wall thin, smooth and colourless (Srivastava and Srivastava, 1976).

It was reported as a parasite in the hyphae of *Saprolegnia ferax,* collected from Gujartal, Kheta Sarai (Distt. Jaunpur) on February 18, 1975 (Srivastava and Srivastava, 1976).

Olpidiopsis ricciae Du Plessis

Trans. Brit. Mycol. Soc. 31 (I&II) : 7-12, 1947.

(Plate 47; Figs. 14 & 15)

Thirumalachar (1947) found it in rhizoids of the species of *Marchantia, Plagiochasma, Fimbriaria, Aneura* and *Notothylus* in Mysore. He reported that a large number of sporangia and oospores were observed within normal and pegged rhizoids. Mature oospores were yellowish brown with a sinous outer wall. The companion cell was mostly lacking. In the rhizoids of *Aneura,* at the place where the young oospores was lodged, the rhizoids turned and twisted forming a nodule like structure indicating definite necrotic effect.

Olpidiopsis saprolegniae var. *indica* Dayal and Thakurji

Sydowia, Annales Mycologici Ser. II., 22(1-4):278-283, 1968

(Plate 48; Figs. 21-25)

Sporangia globose, subglobose, ovoid, ellipsoidal, 18-203 × 15-137 µm, generally terminal, occasionally intercalary in the hypertrophied host filaments; wall thin, smooth, colourless; discharge tubes from 1-3, narrowly cylindrical of variable length; zoospores ovoid or ellipsoidal., 2-5 µm in length; resting spores spherical, 29-74 µm in diam., colourless or somewhat brownish, with a fairly thick wall and covered with protuberances which show remarkable variation from low rounded to long warts, germination not observed; companion cells from 1-3 with each resting spores, ovoid or nearly spherical, 18-37 µm in diameter; wall thin, smooth and colourless (Dayal and Thakurji, 1968 b).

It was found to be parasitic in the filaments of *Saprolegnia* sp. in the Medical College Channel of B.H.U., Varanasi, on 20 January, 1966, culture R.D. 25 (Dayal and Thakurji, 1968 b).

Olpidiopsis saprolegniae var. *levis* Coker

The Saprolegniaceae, p 185, 1923.

(Plate 48; Figs. 14 & 15)

Thallus endobiotic, holocarpic; zoosporangia 1-8 in terminal, occasionally intercalary swelling of the host hyphae; variable in shape from spherical, subspherical, ovoid to elliptical, 25-200 µm in diameter; wall smooth, thin, colourless, discharge tubes 1-5, variable, 0.5-1.5 µm in length, zoospores numerous elliptical, biflagellate, encysted zoospores 1.2-2 µm in diameter; resting spores spherical or ellipsoidal, 25-150 µm in diameter; wall smooth, thick; companion cells 1-2 with each resting spores, spherical to elliptical, smooth-walled, 12.5-60 µm in diameter (Khulbe, 1985).

Parasitic in *Saprolegnia diclina* Humphrey, collected from Nainital lake on October 15, 1982, leg. R.D. Khulbe. The type culture has been deposited in the Department of Botany, Kumaun University, Nainital (0/4).

In India, this species has been found as a natural parasite of *Saprolegnia ferax* (Gruith) Thuret (Khulbe, 1980 a). It could also parasitize *Saprolegnia diclina* Humphrey in artificial inoculation experiment (Khulbe and Bhargava, 1981). Besides, it was capable to parasitize *Isoachlya anisospora* var. *indica*, *I. toruloides*, *Saprolegnia crustosa* var. III, and *S. mixta* var. *asplundii* and unable to infect any species of *Achlya*, *Isoachlya*, *Brevilegnia*, *Aphanomyces*, *Thraustotheca* and *Pythium* (Khulbe and Bhargava, 1981). Gupta and Mehrotra (1988b, 1992) and Rai and Misra (1981) isolated this species on *S. ferax* from Kurukshetra, and Lucknow, respectively. Khare (1992) also found it parasitic on *Saprolegnia ferax* in pond, lake, and river in Bareilly.

Olpidiopsis saprolegniae var. *saprolegniae* (Braun) Cornu

Indian Phytopathology **16** : 271-274, 1963.

(Plate 47; Fig. 4)

Thallus endobiotic, holocarpic; zoosporangia single to many, usually formed in terminal or occasionally intercalary swellings of the host hyphae, mostly spherical, ovoid or ellipsoidal, variable in size, 45-195 μm in diam. (33-45 × 12-20 μm when ellipsoidal); wall smooth, colourless, discharge tubes from 1-3, narrowly cylindrical, 15-24 μm in length; zoospores oval to elongated, 2-4 μm in length, biflagellate; resting spores spherical, 33.5-5.68 μm in diam., exospore wall colourless or brownish, fairly thick, densely covered on outer surface with slender sharp spines, 5-12 μm long; cell contents globular; companion cells 1-2 with each resting spore, usually spherical, 22.5-30 μm in diam.; wall thin, smooth and colourless (Srivastava and Bhargava, 1963).

It was recorded for the first time by Srivastava and Bhargava (1963) as a parasite in the hyphae of *Saprolegnia diclina*, Lal Diggi Pond, Gorakhpur, December 3, 1961.

Olpidiopsis varians *Sha*nor

Indian Phytopathology **17**(3): 249-253, 1964

(Plate 47; Figs. 8-13)

Thallus endobiotic, holocarpic; zoosporangia one to several, usually formed in terminal or occassionally intercalary swellings of the host hyphae, mostly spherical or oval, variable in size, 30-75 μm in diameter; wall smooth, colourless, discharge tubes 1-3, narrowly cylindrical, 12-25 μm in length; zoospores oval to elongate, 3-4.5 μm in length, biflagellate; resting spores spherical, variable in size, 18.5-45 μm in diam. (excluding spines); exospores wall colourless or yellowish, varying from 2.5-10.5 μm in thickness, with a small or large, broad shaped spines or with an entire, undulant or serrate margin; companion cells 1 or 2 with each resting spores, usually spherical, 11-22.5 μm in diam., outer wall continuous with that of the resting spores, usually thick and bearing short ornamentation or smooth, colourless or yellowish (Srivastava, 1964).

It was found parasitic on *Isoachlya anisospora* var. *indica* and *Achlya proliferoides* collected from Asuran-Ka-Pokhra lake, Gorakhpur (Srivastava, 1964). In an host range study, it could parasitize *Achlya orion* and *A. flagellata* and was unable to infect any species of *Saprolegnia, Dictyuchus, Leptolegnia* and *Aphanomyces* (Srivastava, 1966).

Lagenidium Schenk

Verh. Phy. Med. Gesell. Wurzburg, A.F. 9 : 27, 1859.

Aquatic Phycomycetes, P. 982, 1960.

***Lagenidium pygmaeum* Zopf.**

(Plate 48; Figs. 25- 27)

Thallus tubular, irregular with lobes, spherical to ellipsoidal, mostly in a host cell; sporangium bearing broad, short or irregular discharge tube; gametangia similar to sporangia; female gametangia strongly expanded, sometimes with papilla like out growths, 150-300 µm in diameter; oospores spherical, 13-25 µm in diameter, smooth walled, contents with large globule (Gupta and Mehrotra, 1992).

It was collected from the stems of *Myriophyllum spicatum* Linn. (Gupta and Mehrotra, 1992). Karling (1966) reported it from Madras.

Myzocytium Schenk

Uber. Das. Vork. Cont. Zel. im Pflam, p. 10, Wurzburg, 1958.

Aquatic Phycomycetes, p. 973, 1960

***Myzocytium proliferum* Schenk**

(Plate 48; Fig. 28)

Thallus elongate and constricted, consisting of 1-15 segments in a linear series; sporangia many, in bead like chains, hyaline, smooth, spherical, 15-20 µm in diameter; each forming a single narrow cylindrical discharge tube, extending variable distance beyond the host wall, 8-10 µm long by 2-10 µm in diameter; both the gametangia and sporangia are similar in shape and size; oogonia ellipsoidal; oospores spherical, 15-25 µm in diam.; oospore wall smooth, colourless and eccentric, antheridia fusiform (Gupta and Mehrotra, 1992).

In India, this species was found as a parasite of *Spirogyra officinalis* at Hoshiarpur, Punjab by M.S. Randhawa (Chaudhuri, 1931) Later, it was reported by Das Gupta and John (1953) from Lucknow, Manoharachary (1974) from Hyderabad, and by Gupta and Mehrotra (1992) from Kurukshetra.

Plate-47

Olpidiopsis Cornu

Figs. 1-3 : *O. brevispinosa* : 1&2-Zoosporangia showing papillae; 3-zoosporangium, resting spore with companion cell. Scale a (Khulbe, 1980a).

Fig. 4 : *O. saprolegniae* var. *saprolegniae* : Zoosporangia (S) resting spore (R) and companion cell (C). Scale b, (Srivastava & Bhargava, 1963).

Fig. 5 : *O. incrassata* : Zoosporangia, resting spore and companion cell. Scale C, (Srivastava & Srivastava, 1976).

Figs. 6-7 : *O. indica* : Swelling of host hyphae showing zoosporangia (s), resting spore (R) and companion cell (C). Fig. 6 Scale e, Fig. 7 Scale d, (Srivastava, 1975a).

Figs. 8-13 : *O. varians* : 8-Intercalary swelling of the host hypha showing empty zoosporangia, papillae, 9-zoosporangium, resting spore with companion cell, 10-12-resting spores with companion cells. Scale f, (Srivastava, 1964).

Fig. 13 : *O. varians* in *Achlya proliferoides* : Zoosporangia, resting spore with long papillae and companion cell. Scale f. (Srivastava, 1964).

Figs. 14-15 : *O. ricciae* : 14-host cell of thallus showing resting spores; 15-host rhizoids with swelling and resting spore. Fig. 14 - (X 1000), Fig. 15 (X 500), (Thirumalachar, 1947).

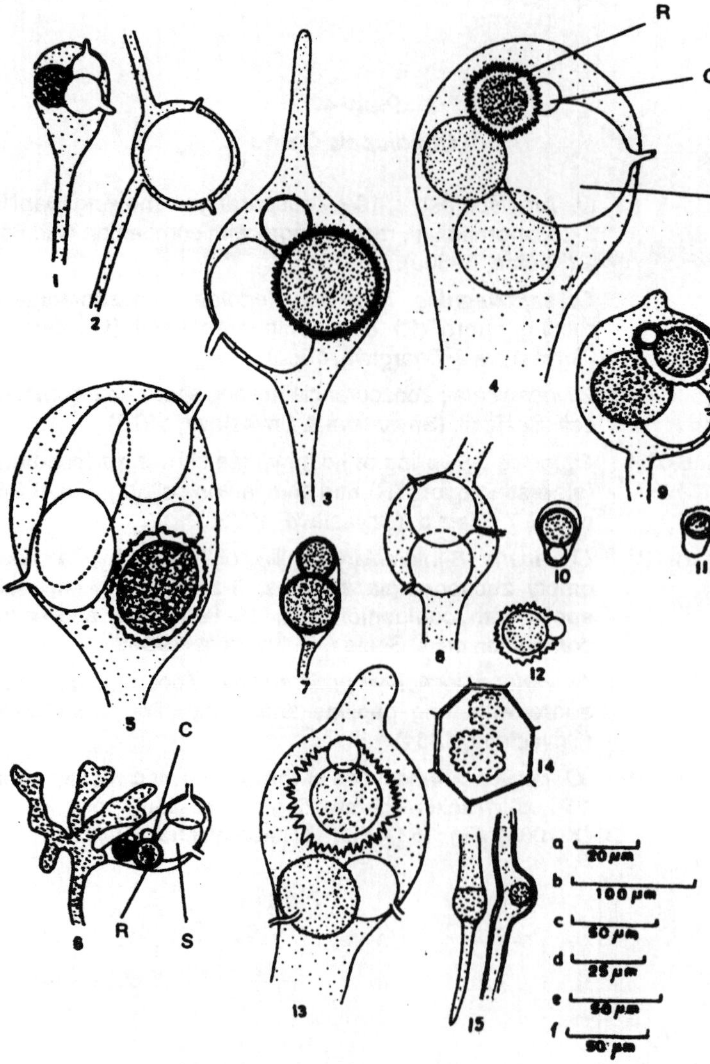
R
C
S
1
2
3
4
5
6
7
8
9
10
11
12
13
14
15
C
R
S
a 20 µm
b 100 µm
c 50 µm
d 25 µm
e 50 µm
f 50 µm

Plate 47

Plate-48

Figs. 1-5 : *Olpidiopsis aphanomyces* : 1-4-Zoosporangia in the host hyphal swelling, 5. resting spore (Kiran & Dayal, 1983).

Figs. 6-9 : *O. achlyae* : Zoosporangia in host apical or intercalary swelling (Khulbe, 1985)

Figs. 10-11: *O. fusiformis* : 10-Sporangium with 2 papillae in apical swelling of the host hypha; 11-resting spore with companion cell in the host hypha (Khulbe, 1980a).

Figs. 12-13 : *O. luxurians* : 12-Zoosporangium in the intercalary swelling of hypha; 13-resting spores with spines. (Khulbe, 1980).

Figs.14-15 : *O. saprolegniae* var. *levis* : 14-Zoosporangium with many long papillae; 15-zoosporangium, resting spore with companion cell in the apical swelling of the host hypha (Khulbe, 1980a).

Figs. 16-17 : *O. pythii* : 16-Zoosporangium in the intercalary swelling of the host hypha, 17-resting spores with fine papillae (Khulbe, 1980a).

Figs. 18-20 : *O. achlyae* : 18-zoosporangia 19-resting spores with small tuberous companion cells. 20- hypha showing sporangium and resting spores with or without companion cells.

(Dayal & Thakurji, 1968b).

Figs.21-25 : *O. saprolegniae* var. *indica* : 21-Sporangia causing gall in a tip of hypha, 22-sporangia sending tubes through wall of sporangium. 23-different sporangia in a hypertrophied tip of *Saprolegnia* sp.

24 : Resting spore with its companion cell and emptied sporangia inside a hypha.

25 : Resting spore merely rough or wasted.

Figs. 21-25- Dayal & Thakurji (1968 b).

Figs. 26&27: *Lagenidium pygmaeum* :

Female and male gametangia in the cells of *Myriophyllum spicatum* (Gupta & Mehrotra, 1992)

Fig. 28 : *Myzocytium proliferum* Schenk : sporangia and discharge tubes (Gupta & Mehrotra, 1992).

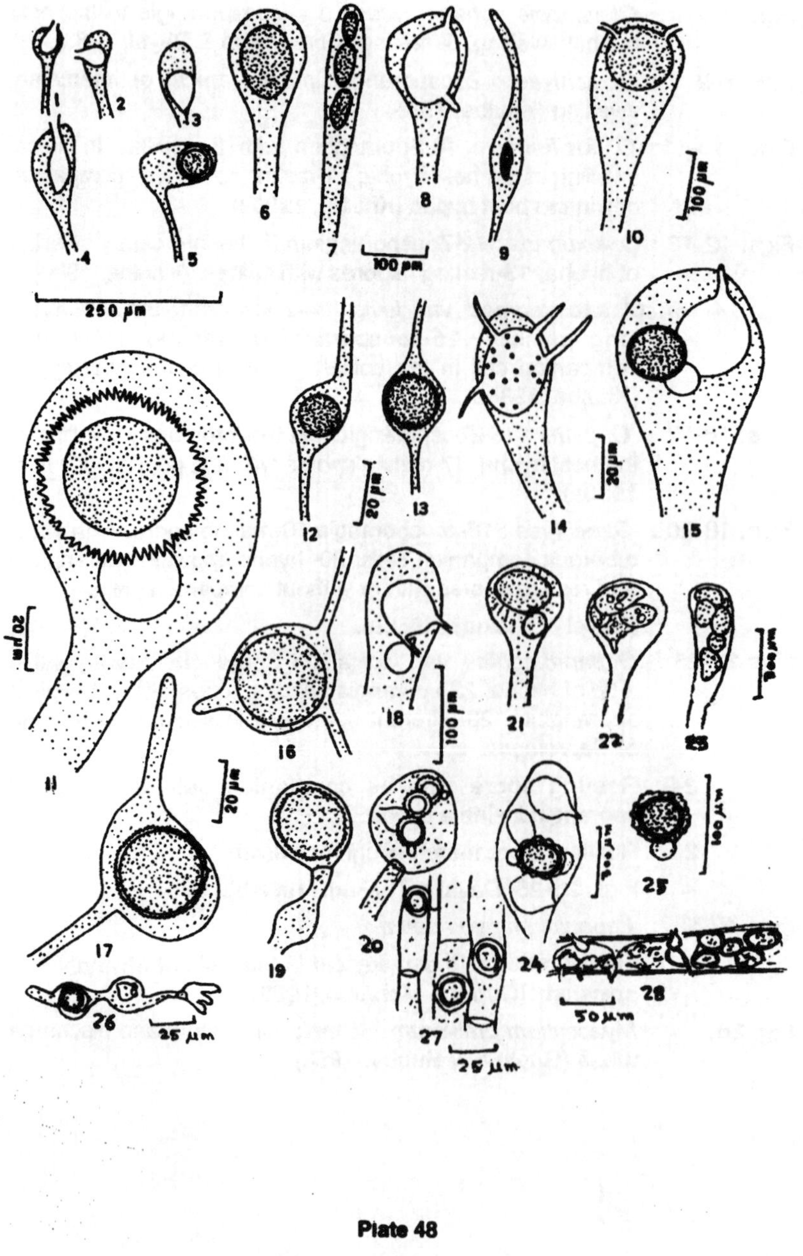
250 µm
100 µm
100 µm
20 µm
20 µm
20 µm
100 µm
20 µm
100 µm
25 µm
25 µm
50 µm

Plate 48

References

Agarwal, G.P., S.K. Hasija and R.P. Mishra (1989). Aquatic fungi from Jabalpur (M.P.) p. 105-112, In Proc. Nat Sem. on Perspectives in Aquatic Biology (Ed. R.D. Khulbe), Papyrus Publishing House, New Delhi, India, pp. 540.

Alabi, R.O. (1971). Seasonal periodicity of Saprolegniaceae at Ibadan, Nigeria. Trans. Brit. Mycol. Soc. *56* : 337-341.

Bhairavnath D. and C. Manoharachary (1985) Mycoecological studies of two water bodies from Andhra Pradesh, India. Indian J. Bot. *8 (1)* : 69-78.

Bhargava, K.S. (1944). A new variety of *Isoachlya anisospora* (de Bary) Coker Current Science **13** : 79.

Bhargava, K.S. (1950). Formation and liberation of zoospores in *Isoachlya anisospora* var. *indica*. The Botanical Gazette *112(1)* : 24-31.

Bhargava, K.S. and B.B. Singh (1965). Observations on Indian Aquatic fungi, III. Some species of Blastocladiales, Monoblepharidales and Peronosporales collected from Gorakhpur. Proc. Nat. Acad. Sci., India. *35* **(B)** : 49-58.

Bhargava, K.S. and G.C. Srivastava (1966). *Isoachlya luxurians* sp. nov. from India. Trans Brit. Mycol. Soc. *49* **(2)** : 269-273.

Bhargava, K.S., K. Swarup and C. S. Singh (1971). Fungi parasitic on certain fresh water fishes of Gorakhpur. Indian Biologist, 3: 65-69.

Bhattacharya, B. and H. K. Baruah (1953). Fungi of Assam. J. University of Gohati, **4** : 287-312.

Butler, E. J. (1907). An account of the genus *Pythium* and some Chytridiaceae. Memb. Dept.. Agri, India. (Bot. Sci) *15* : 1-160.

Butler, E. J. (1911). On *Allomyces,* a new aquatic fungus. Ann. Bot. **25** : 1023-1035.

Butler, E. J. and G.R. Bisby (1931). The fungi of India. Imp. Coun. Agric. Res. India, Sci., Monograph. I *18* : 237 pp.

Chaudhuri, H. (1931). On a *Myzocytium* parasitic on *Spirogyra affinis*. Arkch. Prot. *75* : 472-475.

Chaudhuri, H. (1942). A new genus of the Saprolegniaceae - *Hamidia* gen. nov. Proc. Indian Acad. Sci. *15 (B)* : 225-230.

Chaudhuri, H. and M. L. Banerjee (1942). Indian Watermolds-IV. Proc. Indian Acad. Sci. *15 (B)* : 216-224.

Chaudhuri, H. , and P.L. Kocchar (1935). Indian Watermolds I, Proc. Indian Acad. Sci. **2** *(B)* : 137-154.

Chaudhuri, H. and S.S. Lotus (1936). Indian water moulds II, Proc. Indian, Acad. Sci. *3 (B)* : 328-333.

Chaudhuri, H., P.L. Kochhar, S.S. Lotus, M.L. Banerjee and R.H. Khan (1947). A Hand Book of Indian Water Molds-I. Punjab University, Bot. Publication, Lahore.

Chowdhry, P.N. (1976). Studies on some aquatic fungi with special reference to ecology and taxonomy.Ph. D. Thesis, Rani Durgawati University Jabalpur (M.P.)

Chowdhry, P.N. and G.P. Agarwal (1980a). Studies on distribution of some aquatic fungi in India. Indian Phytopathology *33 (1)* : 107-109.

Chowdhry, P.N. and G.P. Agarwal (1980 b). Studies in seasonal variation on aquatic fungi from Delhi. Indian Phytopathology *33 (4)* : 614-615.

Coker, W.C. (1923). The Saprolegniaceae with notes on other Watermolds. Univ. North Carolina Press-IV, 2000 pp., 63 plates.

Coker, W.C. and V.D. Matthews (1937). Blastocladiales, Monoblepharidales and Saprolegniales, North American Flora *2(1)* : 1-76.

Das-Gupta, S.N. (1982). Discourse on aquatic Phycomycetes of India. Indian Phytopathology *35(2)* : 193-216.

Das-Gupta, S.N. and John, Rachel (Miss) (1953). Studies in the Indian Aquatic fungi. 1. Some Watermold of Lucknow. Proc. Indian Acad. Sci., *38 (4)* **B** : 165-170.

Das-Gupta, S.N. and R. John (1988). A contribution to our knowledge of the genus *Blastocladia*. Indian Phytopathology *41(4)* : 521-547.

Dayal, R. (1958). Some Aquatic fungi of Allahabad. A Taxomomic study. Proc. Nat. Acad. Sci. (India), Allahabad. *28(B) (1)*: 49-57.

Dayal, R. (1961d).. Development of zoosporangium and liberation of zoospores in *Achlya aplanes* Maurizio var. *indica* Sak. et Dayal. Cytologia *26(3-4)* : 372-377.

Dayal, R. (1968a). Studies in aquatic fungi of Varanasi-IV. New records and notes on taxonomy. Hydrobiologia *32* : 237-244.

Dayal, R. and Usha Kiran (1985). Zoosporic Fungi of India, Inter-India Publication, New Delhi, pp. 297.

Dayal, R. and R.N. Tandon (1962). Ecological studies of some aquatic Phycomycetes. Hydrobiologia *20* : 121-127.

Dayal, R. and R.N. Tandon (1963). Ecological studies of some aquatic Phycomycetes-II. Fungi in relation to chemical factors of the water. Hydrobiologia 22 : 324-330.

Dayal, R. and Thakurji (1965). Studies in aquatic fungi of Varanasi. Proc. Nat. Acad. Sci., India *35 (B)* : 309-319.

Dayal, R. and Thakurji (1966). The occurrence and distribution of aquatic fungi in certain ponds of Varanasi. Hydrobiologia 27 : 548-558.

Dayal, R. and Thakurji (1968a). Studies on aquatic fungi of Varanasi-V. A taxonomic study. Proc. Nat. Acad. Sci., India. *38 B* : 32-38.

Dayal, R. and Thakurji (1968b). Studies in aquatic fungi of Varanasi III, observation on some parasitic aquatic Phycomycetes. Sydowia Annales Mycologici Ser.II, *22(1-4)* : 278-283.

Dayal R. and Thakurji (1969). Studies in Aquatic Fungi of Varanasi, II. A new species of *Achlya* with notes on other species. Mycopathologia et. Mycologia Applicata *38* (1-2) : 169-174.

Dayal, R. and Usha kiran (1978). *Catenomyces persicinus* - a new record from India. Indian Phytopathology. *31* : 266-268.

Dick, M.W. (1969). The *Scoliolegnia asterophora* aggregate formerly *Saprolegnia asterophora* de Bary. Bot. J. linn. Soc. *62* : 255-266.

Dick, M.W. and H.V. Newby (1961). The occurrence and distribution of Saprolegniaceae in certain soils of S.E. England. I-Occurrence. J. Ecol. *49* : 403-419.

Dube, B. (1982). An ecological study of aquatic fungal flora of lakes of Bhopal. Ph. D.. Thesis, Bhopal University, Bhopal.

Emerson, R. (1941). An experimental study of the life cycles and taxonomy of *Allomyces* Lloydia 4 : 77-144.

Forbes, E.J. (1935). Watermolds of Manchester District. Mem. Manchr. Let. Phil. Soc., *79* : 1-11.

Gupta, A.K. and R.S. Mehrotra (1987). Occurrence of watermolds in relation to Hydrogen ion concentriation in tanks of Kurukshetra, India. Biome *2(1)* : 1-7.

Gupta, A.K. and R.S. Mehrotra (1988 a) Watermolds from Kurukshetra (India) Saprolegniales. I.B.C. *5* : 145-148.

Gupta, A.K. and R.S. Mehrotra (1988 b) Parasitic Watermolds from Kurukshetra, India. Advances in Plant Sciences *1* (2 suppl.) : 373-380.

Gupta, A.K. and R.S. Mehrotra (1989 a). Seasonal periodicity of aquatic fungi in tanks at Kurukshetra, India. Hydrobioligia *173* : 219-229.

Gupta, A.K. and R.S. Mehrotra (1989 b). Occurrence, distribution and seasonal variation of watermolds as affected by chemical factors in tanks of Kurukshetra-India. Proc. Nat. Acad. Sci., India *59(B)*, III : 303-313.

Gupta, A.K. and R.S. Mehrotra (1992). Some fresh watermolds from Kurukshetra-Lagenidiales and blastocladiales. Advances in Plant Sciences *5(1)* : 116-123.

Hamid, A. (1942). Indian watermolds III. Proc. Indian Acad. Sci. *15(4) B* : 206-215.

Hasija, S.K. and S. Batra (1978). The distribution of *Achlya americana* (Saprolegniales) in different aquatic habitats at Jabalpur, India. Hydrobiologia *61 (3)* : 277-279.

Hasija, S.K. and M.A. Khan (1983). Some new records of aquatic fungi from Jabalpur, India. Proc. Nat. Acad. Sci., India. *53(B)*, I : 71-72.

Hasija, S.K. and M.A. Khan (1987). Seasonal occurrence and distribution of Aquatic fungi in relation to the water quality of Jabalpur lakes. Perspectives in Mycological Research-1 Prof. G.P. Agarwal Festschrift Volume : *79-87*. Today & Tomorrow, New Delhi.

Hohnk, W. (1932). A new parasitic *Pythium*. Mycologia. *24* : 489-507.

Hughes, G.C. (1962). Seasonal periodicity of the Saprolegniaceae in the South Eastern United States.. Trans. Brit. Mycol. Soc. *45* : 519-531.

Indira, G., R.P. Mishra and G.P. Agarwal (988). Some fungi from the Supatal lake at Jabalpur (M.P.). Nat. Acad. Sci., letters *11(3)* : 65-67.

Ito, S. and Nagai, M. (1931). On the root disease of the seeds and seedlings of rice plant caused by some aquatic fungi. J. Fac. Agric. Hokkaido Imp. Univ., 32 : 45-69.

Johnson, T.W. Jr. (1956). The genus *Achlya;* morphology and taxonomy. Univ. of Michigan Press, Ann Arbor. 180 pp.

Johnson, T.W. Jr. (1971). Aquatic fungi of Iceland-*Pythium*. Mycologia 63 : 517-536.

Kapadnis, S.D. (1980). Aquatic Fungi of Maharashtra-A list., University of Poona, Personal Communication.

Karling, J.S. (1966). The *Chytrids* of India with a supplement of other zoosporic fungi. Sydowia Annales Mycologici Ser. II. *6* : 1-125.

Khan, M.A. (1981). Studies on the Ecology of some aquatic fungi. Ph. D. Thesis, University of Jabalpur, Jabalpur, 166 pp.

Khare, A.K. (1992) Taxo-ecological studies on watermolds of Bareilly. Ph. D. Thesis, Ruhelkhand University, Bareilly, pp. 168.

Khulbe, R.D. (1977). Taxonomic and Ecological studies of watermolds in Naini Tal and its suburbs. Ph.. D. Thesis. Agra University, Agra. pp. 113.

Khulbe, R.D. (1980 a). Occurrence of parasitic watermolds in some lakes of Nainital, India. Hydrobiologia, *70* : 119-121.

Khulbe, R.D. (1980 b). Occurrence of watermolds in some lakes of Nainital, India. Hydrobiologia *74* : 77-80.

Khulbe, R.D. (1980 c). Occurrence of watermolds in relation to Hydrogen ion concertration in some lakes of Nainital, India. Hydrobiologia 69 : 2-5.

Khulbe, R.D. (1982). Distribution of aquatic fungi in relation to some ecological factors. Geobios *8* : 214-216.

Khulbe, R.D. (1982). *Geolegnia inflata* Coker & Harvey-a new record for Indian aquatic fungi from Nainital, India. Geobios New Report *1* : 35-36.

Khulbe, R.D. (1983 a). *Protoachlya oryzae* sp. nov. from Naini Tal. Current Science 52 (?) : 428.

Khulbe, R.D. (1983 b). Aquatic fungal flora in rhizosphere and non rhizosphere soils in some cultivated fields of Naini Tal. Science and Culture, *49* : 105-106.

Khulbe, R.D. (1983 c). *Scoliolegnia centrica* -a new watermold from India. Sydowia, Annales Mycologici Ser. II. 36 : 150-153.

Khulbe, R.D. (1985). Studies in Aquatic Phycomycetes of Nainital in relation to Taxonomy, Physiology, Ecology and Pathology. D.Sc. Thesis, Kumaun University, Nainital. pp. 335.

Khulbe, R.D. (1987). Some interesting watermolds and their adaptability in Kumaun Himalaya pp. 53-62, in Perspectives in Mycological Research. Prof. G.P. Agarwal Festschrift Volume I. Today & Tomorrow, New Delhi.

Khulbe, R.D. (1991). An ecological study of watermolds of forest soils of Kumaun Himalaya, India. Tropical Ecology *32(1)* : 127-135.

Khulbe, R.D. and B.L. Verma (1983). Aquatic fungi in some rice fields of Tarai, Nainital. Trop. Plant Sci. Res. *1(3)*: 269-270.

Khulbe, R.D. and S.C. Sati (1983). A new species of *Dictyuchus* and key to the other species. Bibliotheca Mycologica *91* : 609-617.

Khulbe, R.D. and B.L. Verma (1987). *Saprolegnia bhargavii* sp. nov. from India Nova Hedwigia *44(1-2)* : 215-219.

Khulbe, R.D., B.L. Verma and D.L. Verma (1983). A new medium of Saprolegniaceae. Bibliotheca Mycologica *91* : 557-561.

Khulbe, R.D. and K.S. Bhargava (1977). Distribution and seasonal periodicity of watermolds in some lakes in Nainital Hills, India. Hydrobiologia *54(1)* : 67-72.

Khulbe, R.D. and K.S. Bhargava (1981). Studies in parasitic watermolds of Kumaun Himalaya; The host range of five species of *Olpidiopsis*. Hydrobiologia *76* : 45-47.

Khulbe, R.D. and K.S. Bhargava (1983). Frequency of watermolds in relation to nitrate, sulphate and phosphate in some lakes of Naini Tal, India. Tropical Ecology *24(2)* : 180-187.

Khulbe, R.D. and S.C. Sati (1979). A new record of watermolds from Nainital. India. Geobios *6* : 229.

Khulbe, R.D. and S.C. Sati (1983). A new species of *Dictyuchus* and key to the other species. Bibliotheca Mycologica *91:* 609-617.

Kiran, U. and R. Dayal (1983). Fresh watermolds from Varanasi (India). VI. Some Oomycetes. Proc. Nat. Acad. Sci., India *53(B)* : 50-52.

Kiran, U. and R. Dayal (1992). A new host record of *Blastocladia* and *Gonapodya* Proc. Nat. Acad. Sci. India *62(B)* III : 503-506.

Lacy, R.C. (1949). Studies in aquatic Phycomycetes. Indian Phytopathology *2* : 134-141.

Lacy, R.C. (1955). Studies in aquatic Phycomycetes II. Indian Phytopathology *8* : 208-209.

Madhusudan, Rao, M. (1981). Studies on the myco-ecology of certain aquatic, semi-aquatic and terrestrial ecosystems of Mannanur forest Andhra Pradesh, India. Ph.D. Thesis, Osmania Univ. Hyderabad.

Madhusudan, Rao, M. C. and C. Manoharachary (1980). Taxo-ecological studies on some noteworthy fungi. Geobios 7 : 165-166.

Manoharachary, C. (1973). A note on some aquatic Phycomycetes from Hyderabad (India). Current Science *42 (24)* : 874.

Manoharachary, C. (1974). A note on some interesting fungi from Hyderabad (India). Current Science *43(4)* : 129-131.

Manoharachary, C. (1974). Investigation of the microbial ecology of two fresh water ponds, muds and soils of Hyderabad district, Andhra Pradesh, India. Ph.D. Thesis, Osmania Univ., Hyderabad.

Manoharachary, C. (1976). Occurrence of some fungi in pond waters and mud of Hyderabad, India. Geobios 3 : 196-197.

Manoharachary, C. (1977). Studies on Ecological behaviour of certain fungi. M.V.M. Patrika *12(1)* : 17-18.

Manoharachary, C. (1979 a). Ecological studies on some species of *Allomyces* from Andhra Pradesh, India. Nova Hedwigia *63* : 179-182.

Manoharachary, C. (1979 b). Taxo-ecological studies on some freshwater pond mud fungi of Hyderabad, India. Nova Hedwigia *63* : 197-207.

Manoharachary, C. (1981). The taxonomy and ecology of freshwater Phycomycetes from India. Indian Rev. Life Sci. *1* : 3-21.

Manoharachary, C. and R. Rao (1981). Seasonal variation and distribution of fungi in two freshwater ponds of Andhra Pradesh, India. Proc. Indian Aad. Sci. *90(3)* : 237-243.

Manoharachary, C., D. Bhairavanath and M. Madhusudan Rao (1983). Taxo-ecological studies on some aquatic fungi from India. Bibliotheca Mycologica *91* : 457-462.

Mer, G.S. and R.D. Khuble (1984). Watermolds of Sat Tal (Nainital), India. Sydowia, Annales Mycologici Ser. II. *37* : 195-207.

Mer, G.S. and S.C. Sati (1987). An unusual strain of *Apodachlya minima* Coker and Leitner. Current Science *56(17)* : 906-907.

Mer, G.S., S.C. Sati and R.D. Khulbe (1980). Occurrence, distribution and seasonal periodicity of some aquatic fungi of Sat Tal (Nainital) India. Hydrobiologia *76* : 200-205.

Mer, G.S. (1982). Ph.D. Thesis, Kumaun University, Naini Tal.

Mer, G.S., and R.D. Khulbe (1981). An addition to Indian aquatic fungi. Indian Phytopathology *34* : 387-388.

Middleton, J.T. (1943). The taxonomy, host range and geographical distribution of the genus *Pythium*. Mem. Torrey Bot. Club. Mem. *20* : 1-171.

Milovtsova, M.O. (1935). Vadiah gribi Kharkov taiago okelits (Aquatic Phycomycetes from Kharkov and its environs) kharkiv k: Univ Trudi N-D-Inst Bot. I : 28-37.

Misra, J.K. (1980). Ph. D. Thesis, Lucknow University, Lucknow.

Misra, J.K. (1982). Occurrence, distribution and seasonality of aquatic fungi as affected by chemical factors in six alkaline ponds of India. Hydrobiologia *97*: 185-191.

Misra, J.K. (1983). Bibliotheca Mycologica 91 : 425-455.

Misra, R.C. and R.S. Dwivedi (1987). Aquatic Moulds from Gujar lake, Jaunpur-II. J. Indian Bot. Soc. *66* : 203-208.

Mishra, R.P., S.K. Hasija and G.P. Agarwal (1990). Aquatic fungi from Ganga Sagar Lake, Jabalpur. Proc. Nat Acad. Sci., India. *59 B(III)* : 351-355.

Mitra, A. (1935). Studies on Indian aquatic fungi. Proc. Indian Science, Congress 22 : 252.

Naumov, N.A. (1954). Flora gribor leningradskoi, oblasti, Vypusk 1-A rkhimistesely, Ficomistsety (flora of the fungi Leningrad region : Part I. Archimycetes and Phycomycetes) Moscow and Leningrad : Academy of Science, USSR.

Patil, S.D. and B.P. Kapadnis (1977). Aquatic Phycomycetes of Maharashtra. M.V.M. Patrika 12 (I) : 13-16.

Pavgi, M.S. and S.L. Singh (1969). *Aphanomyces* root rot of cauliflower in Varanasi, a new record for India. Experientia 25 : 989.

Perrott, P.E. (1960). The ecology of some aquatic Phycomycetes. Trans. Brit. Mycol. Soc. **43** : 19-30.

Prabhuji, S.K. (1979). Studies on some lower fungi occurring in certain soils of Gorakhpur. Ph. D. Thesis, Gorakhpur University, Gorakhpur.

Prabhuji, S.K. (1984). Studies on some Water Moulds occurring in certain soils of Gorakhpur. J. Indian Bot. Soc. **63** : 387-396.

Prabhuji, S.K., G.C. Srivastava and S.K. Sinha (1984) Aquatic fungi of India, IV, *Brevilegnia* Coker and Couch, Kavaka, **12** : 99-106.

Prabhuji, S.K. and S. K. Sinha (1993) Two new watermolds from India : *Allomyces recurvus* sp. nov. and *Brevilegnia indica* sp. nov. Neo Botanica **1 (1&2)** : 31-38.

Prabhuji, S.K. and G.C. Srivastava (1977 a). Some members of Blastocladiales and Peronosporales occurring in certain soils of Gorakhpur. Geobios **4** : 154-155.

Prabhuji, S.K. and G.C. Srivastava (1977 b). Some members of Saprolegniaceae occurring in the soils of Gorakhpur. Geobios **4** : 258-259.

Prabhuji, S.K. and G.C. Srivastava (1978 a). Addition to lower fungi of India. Geobios *5* : 35-36.

Prabhuji, S.K. and G.C. Srivastava (1978 b). Further additions to lower fungi of India. Geobios *5* : 137-138.

Rai, J.N. and J.K. Misra (1976). *Achlya hypogyna* - A notable addition to Indian aquatic fungi. Current Science *45(23)* : 843.

Rai, J.N. and J.K. Misra (1977 a). A new variety of *Achlya stellata* de Bary. Current Science, *46* : 28.

Rai, J.N. J.K. Misra (1977 b). Aquatic fungi from alkaline ponds and soils. Kavaka **5** : 73-78.

Rai, J.N. and J.K. Misra (1981). *Olpidiopsis* : A parasitic fungus from alkaline water. Geophytology *11* : 96-98.

Rao,M. and C. Manoharachary (1983). Ecology of fresh water *Phycomycetes* from Andhra Pradesh, India. Indian Phytopathology *36* : 431-437.

Roberts, R. E. (1963), A study of the distribution of certain members of the Saprolegniaceae. Trans. Brit. Mycol. Soc. **45** : 213-224.

Saksena, R.K. and K.S. Bhargava (1944). A new variety of *Isoachlya anisospora* (de Bary) Coker. Current Science *13* : 79.

Saksena, S.B. and C. Rajgopalan (1958). Studies on aquatic fungi of Saugar. Jour. Univer. of Saugar **7** : 7-20.

Sati, S. C. (1981). Aquatic fungi of Kumaun in relation to fish infection. Ph.D. Thesis. Kumaun University, Naini Tal. pp. 111.

Sati,S.C. and R.D. Khulbe (1980). Occurrence of aquatic fungi in some soils of Naini Tal Hills. Geobios **7** : 42-43.

Sati,S.C. and R.D. Khulbe (1981). *Protoachlya paradoxa* coker - a new record for Indian aquatic fungi. Current Science **50** : 740.

Sati,S.C. and G.S. Mer (1988). Fungal flora of Kumaun region : Aquatic Mastigomycotina. Indian Phytopathology *41(1)* : 39-43.

Scott, W.W. (1961). A monograph of the genus *Aphanomyces*. Tech. Bull. 15. Va. sgr. Exp. Sta. 95 pp. Va, Polytechnic Institute Blacksburg.

Seymour, R.L. (1970). The genus *Saprolegnia*. Nova Hedwigia Z. Kryptogamemkd, **19** : 1-124.

Shanor, L. (1940) Studies in the genus *Olpidiopsis* III. Some observation on the host range of certain species. J. Elisha Mitchel Sci. Soc. **56** : 165-176.

Singh, S.L. and M.S. Pavgi, (1977). *Aphanomyces* root rot of cauliflower. Mycopathologia *61* : 167-172.

Sparrow, F.K. Jr. (1929). A note on the occurrence of two rotifer - Capturing Phycomycetes, Mycologia **21** : 90-96.

Sparrow, F.K. (1943). Aquatic Phycomycetes exclusive of the Saprolegniaceae and *Pythium*. Univ. of Michigan press, Ann Arbor. pp. 785.

Sparrow, F.K. (1960). Aquatic Phycomycetes (2nd ed.). University of Michigan press. Ann Arbor. 1187 pp.

Sridhar; K.R. and K.M. Kaveriappa, (1982). Aquatic fungi of the Western Ghat forest in Karnataka. Indian Phytopathology, *35* : 293-296.

Srivastava, G.C. (1964). Observation on Indian aquatic fungi II. Occurrence of *O. varians* Shanor at Gorakhpur. Indian Phytopathology 17 *(3)* : 249-253.

Srivastava, G.C. (1966). The host range of four species of *Olpidiopsis*. Trans. Brit. Mycol. Soc. *49* : 69-72.

Srivastava, G.C. (1967 a). Some new species of Saprolegniaceae collected from Gorakhpur, India. Hydrobiologia *30(2)* : 281-292.

Srivastava, G.C. (1967 b). Ecological studies on some aquatic fungi of Gorakhpur, India. Hydrobiologia *30* : 385-404.

Srivastava, G.C. (1975 a). *Olpidiopsis indica* sp. nov. from India. Current Science **44** : 642-644.

Srivastava, G.C. (1975 b). *Aphanomyces stellatus* de Bary - a new record from India. Geobios **2** : 192.

Srivastava, G.C. (1975 c). *Achlya stellata* de Bary - a new record from India. Current Science *44(4)* : 139.

Srivastava, G.C. (1982). The host range of *Olpidiopsis indica* Srivastava. J. Indian Bot. Soc. *61* : 108-110.

Srivastava, G.C. and K.S. Bhargava (1963). Observations on Indian aquatic fungi I. Three species of *Olpidiopsis* from Gorakhpur. Indian Phytopathology *16* : 271-274.

Srivastava, G.C. and S. Mekrani (1980). *Hypophthalmichthys molitrix* new host of *Dictyuchus sterile*. Geobios **7** : 28-29.

Srivastava, G.C., and R.C. Srivastava (1976). *Olpidiopsis incrassata* - A new record from India. Indian Phytopathology *29(1)* : 83-84.

Srivastava, G.C. and R.C. Srivastava (1977). *Achlya caroliniana* Coker - a new record from India. Current Science *46(12)* : 422.

Srivastava, R.C. (1981). *Dictyuchus anomalus* Nagai - A new record for India. Biovigyanam **7** : 81-82.

Srivastava, R.C. (1985). The genus *Saprolegnia* C.G. Nees in India. J. Econ., Tax., Bot., Vol. *7(1)* : 145-151.

Thakurji, (1967). Some new records of Indian aquatic fungi of Varanasi. Indian Phytopathology ***20*** : 385-386.

Thakurji, (1970). Studies on aquatic fungi of Varanasi X - additional new interesting records of aquatic Phycomycetes. Hydrobiologia ***36*** : 179-186.

Subramanium, C.V. (1952). Fungi isolated and recorded from Indian Soils, Jour. Madras Univ. **22** : 206-222.

Thirumalachar, M.J. (1947). Some fungal disease of Bryophytes in Mysore. Trans. Brit. Mycol. Soc., ***31*** : 7-12.

Verma, B.L. (1984).. Studies on lower Phycomycetes in some agricultural fields of Tarai, Nainital with special reference to Saprolegniaceae and Pythiaceae. Ph.D. Thesis, Kumaun Univ, Nainital, p. 139.

Verma, B.L. (1987). Zoosporic fungi in wheat fields of Kumaun Himalaya. Madras Agricultural Journal ***74(1)*** : 1-5.

Verma, B.L. and R.D. Khulbe (1983 a). *Achlya oryzae* Nagai, a new record for Indian aquatic fungi. Current Science ***52(5)*** : 218-219.

Waterhouse, G.M. (1942). Some watermolds of the Hogsmill river collected from 1937-1939. Trans. Brit. Mycol. Soc. **25** : 315-325.

Subject Index

Author Index